KEY CONCEPTS IN

ECONOMIC GEOGRAPHY

The *Key Concepts in Human Geography* series is intended to provide a set of companion texts for the core fields of the discipline. To date, students and academics have been relatively poorly served with regards to detailed discussions of the key *concepts* that geographers use to think about and understand the world. Dictionary entries are usually terse and restricted in their depth of explanation. Student textbooks tend to provide broad overviews of particular topics or the philosophy of Human Geography, but rarely provide a detailed overview of particular concepts, their premises, development over time and empirical use. Research monographs most often focus on particular issues and a limited number of concepts at a very advanced level, so do not offer an expansive and accessible overview of the variety of concepts in use within a subdiscipline.

The *Key Concepts in Human Geography* series seeks to fill this gap, providing detailed description and discussion of the concepts that are at the heart of theoretical and empirical research in contemporary Human Geography. Each book consists of an introductory chapter that outlines the major conceptual developments over time along with approximately twenty-five entries on the core concepts that constitute the theoretical toolkit of geographers working within a specific subdiscipline. Each entry provides a detailed explanation of the concept, outlining contested definitions and approaches, the evolution of how the concept has been used to understand particular geographic phenomena, and suggested further reading. In so doing, each book constitutes an invaluable companion guide to geographers grappling with how to research, understand and explain the world we inhabit.

Rob Kitchin
Series Editor

KEY CONCEPTS IN
ECONOMIC
GEOGRAPHY

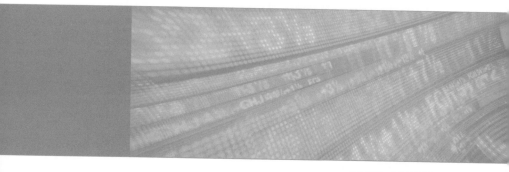

YUKO AOYAMA
JAMES T. MURPHY
SUSAN HANSON

Los Angeles | London | New Delhi
Singapore | Washington DC

SAGE Publications Ltd
1 Oliver's Yard
55 City Road
London EC1Y 1SP

SAGE Publications Inc.
2455 Teller Road
Thousand Oaks, California 91320

SAGE Publications India Pvt Ltd
B 1/I 1 Mohan Cooperative Industrial Area
Mathura Road, Post Bag 7
New Delhi 110 044

SAGE Publications Asia-Pacific Pte Ltd
33 Pekin Street #02-01
Far East Square
Singapore 048763

Library of Congress Control Number: 2010925495

British Library Cataloguing in Publication data

A catalogue record for this book is available from the British Library

ISBN 978-1-84787-894-6
ISBN 978-1-84787-895-3 (pbk)

Typeset by C&M Digitals (P) Ltd, Chennai, India
Printed by CPI Antony Rowe, Chippenham, Wiltshire
Printed on paper from sustainable resources

CONTENTS

Contents

LIST OF FIGURES

ACKNOWLEDGEMENTS

We thank Rory Horner and Joe Pierce, both at Clark University's Graduate School of Geography, for providing excellent research assistance.

We also thank Marie Anselm (Clark '10) for help in indexing the book.

INTRODUCTION

What is economic geography?

What are the major factors that explain the recent growth of the Chinese economy and the relative decline of the United States economy? What explains persistent poverty in pockets of global cities such as New York, London and Tokyo, and what prompted the emergence of vast urban slums in Calcutta? What are the impacts of globalization on people's jobs and livelihoods in different parts of the world? Explaining the causes and consequences of uneven development within and between regions is a central concern for economic geographers. The discipline's goal has long been to offer multi-faceted explanations for economic processes – growth and prosperity as well as crises and decline – manifested across territories at various scales: local, regional, national and global. Contemporary economic geographers study geographically specific factors that shape economic processes and identify key agents (such as firms, labour and the state) and drivers (such as innovation, institutions, entrepreneurship and accessibility) that prompt uneven territorial development and change (such as industrial clusters, regional disparities and core – periphery).

Over its history, economic geographers have considered various key geographically specific endowments as drivers of territorial development. In the early days of the subdiscipline, the economy was dominated by agriculture, and therefore climate and natural-resource endowments mattered significantly, as did labour supply. As industrialization advanced during the twentieth century, the focus shifted to the geography of firms and industries, factory wages, production processes, technology and innovation, the quality and skills of labour, and the role of the state in inducing and promoting industrialization. Most recently, emphasis has shifted away from geographically specific resource endowments in forms of tangible and quantifiable indicators towards research that focuses on often unquantifiable and intangible contributions to territorial development, and, in particular, social endowments such as institutions, networks, knowledge and culture. Analyses that recognize differences among agents have emerged, taking into account, among other things, race, class and gender. New research themes are also emerging, ones that focus for example on financialization, consumption, the knowledge economy and sustainable development.

Still, economic geography as a subdiscipline of geography has frequently been misunderstood. The confusion is partly attributed to its multiple roots, its heterodox methodologies and its various overlapping interests with other social science disciplines. As we shall show, contemporary economic geography has a long history and represents a wide range of inter- and intra-disciplinary influences from fields such as geographical economics, regional science, urban/regional studies, regional economic development planning and economic sociology. Despite the complex and diverse web of influences, economic geographers are bound together by a common interest in economic processes as fundamentally territorial processes as manifested in the economic fates of communities, cities and regions. The goal of the subdiscipline is to understand the economic processes of a place, rather than using economic factors as independent or intervening variables to explain environmental change (cultural ecologists) or socio-cultural change (cultural/social geographers).

2 Origins and evolution

Various interpretations exist on the origin and historical lineage of economic geography. Some argue that the earliest roots of economic geography were deeply linked to British colonialism, which necessitated the study of commercial geography to better understand and improve trade routes and modes of transportation (see Barnes, 2000). Others point to the Germanic location theories of Heinrich Von Thünen and Alfred Weber (who were then followed by Walter Christaller and August Lösch) as the roots of economic geography. Their goal was to develop optimal location patterns for the most efficient functioning of farms, factories and cities, given geographical endowments and accessibility (e.g. transport costs). Location modelling subsequently crossed the Atlantic, where it was incorporated into North American economic geography and became an important foundation of regional science, thanks to Walter Isard.

Yet another lineage of economic geography comes from Alfred Marshall, a notable British economist who was a central figure in the marginal revolution that transformed economics in the early twentieth century. Marshall was among the first to articulate the phenomenon of industrial agglomerations and he highlighted the importance of economies of scale (i.e. the sharing of labour pools and infrastructure) in industrialization. Research on agglomerations/clusters continues to occupy centre stage in contemporary economic geography, although today attention is increasingly shifting from economic to social, cultural and institutional dimensions of clusters. Finally, lineage is also traced to North American human-environmental geography, in which appropriate use of abundant extractive resources in a territory preoccupied scholars and policy makers alike. For example, when the journal *Economic Geography* was established at Clark University in 1925, Wallace W. Atwood, the founding editor, intended to cover areas of research that engaged with human adaptations to natural resources in the process of industrialization. He wrote:

> As citizens of the modern commonwealth we are prepared to act intelligently on the larger problems of national and international economic and social relationships only when we appreciate the possibilities and limitations of the habitable portions of the earth. (Atwood, 1925)

The topics covered were mostly natural-resource industries, such as timber, coal, wheat and the grain trade, and included the study of population in relation to cultivated land.[1]

Thus, economic geography since its inception combined multiple trajectories involving multiple, and sometimes contradictory, epistemologies and methodologies. For example, those who followed Germanic and Marshallian traditions used deductive, scientific methods seeking abstraction and universal application, while those in North American human-environmental traditions (e.g. Keasbey, 1901a; 1901b; Smith, 1907; 1913) regarded economic geography as a 'descriptive science' that would contribute to knowledge by gathering 'concrete information' about lands (i.e. natural resources) and their usefulness to humanity. The influence of these early papers can be seen in the work of Hartshorne's (1939) regionalist approach, which is largely empirical and idiographic, and they maintain strong ties to physical geography (e.g. see Huntington, 1940).

Subsequently, the Great Depression and the catastrophe of World War II dramatically changed economic geography as imperial expansion ended and the extreme consequences of implementing deterministic concepts into policy and then practice became evident (e.g. the Holocaust). As Fisher (1948: 73) observed, economic geography had 'become an apology for the *status quo*'. This led to a few developments during the 1950s and 1960s. For one, economic geographers sought to move the subdiscipline away from descriptive studies and deterministic theories. For another, the idiographic tradition that emphasizes the specificity of places and regions came head-to-head in the post-World War II period with the theoretically driven views of both neo-classical economics and Marxist structuralism. The fact that some of the early work in human-environment economic geography – much like the discipline of geography itself – was plagued by environmental determinism did not help support their strictly descriptive approach.

Finally, this period was most profoundly marked by the shift towards spatial science in economic geography, a transformation that was inspired in large part by the work of German location theorists such as Weber, Christaller and Lösch. Otherwise known as the 'space cadets', quantitative, theoretical economic geographers and regional scientists sought to develop universal, abstract, and explanatory spatial theories for industrial location and regional economic evolution. Through works by Hoover (1948), Isard (1949, 1953), and Berry and Garrison (1958), the quantitative revolution was launched, eventually transforming all areas of human geography from descriptive analysis focused on specificity to scientific analysis aimed at developing generalized theories to explain geographical phenomena. The development of new areas of research, including transportation geography, innovation diffusion studies and behavioural geography, further challenged the old methodological orientation. Research on urban economics with abstract models of transportation, accessibility and the structure of metropolitan areas (e.g. Alonso, 1964) dominated the field, and this trend remained strong throughout the 1970s, as indicated by the dominance of published articles in *Economic Geography*. At its apex (e.g. see Scott, 1969; Chojnicki, 1970), the analytical methods used (e.g. Monte Carlo simulations, equilibrium analyses, and predictive statistics) were complex, often highly abstract, and required extensive mathematical training.

Diverging trajectories: 1970s–1990s

Beginning the late 1960s, some economic geographers began making an initial foray into new areas of scholarship. This was largely in response to geo-political conflicts, environmental and political crises, the social turmoil of the late 1960s, and the global economic slow-down of the 1970s. The limits of American scientific management and associated mass production, the Keynesian welfare state, and deindustrialization in United Kingdom and the United States not only helped to generate new themes for economic geographers but also prompted stronger links to macro-theories on the structure and organisation of economies. David Harvey's conversion from positivism (e.g. see Harvey, 1968, 1969) to Marxism and political activism (e.g. see Harvey, 1974) is emblematic of the increased interest in the early 1970s in politics, as well as poverty, race and class, especially in the inner cities of the industrialized world. These changes also opened economic geography to the influences of structuralism, such as dependency theory (Frank, 1966) and world-system theory (Wallerstein, 1974), then dominant in other social sciences.

Others chose to learn from work primarily by labour economists in North America (Bluestone and Harrison, 1982; Piore and Sabel, 1984) and in France (Aglietta, 1979), or sought to engage with newly emerging branches of economics (evolutionary economics by Nelson and Winter, 1974, institutional economics by Williamson, 1981) and business/ management studies (e.g. work on industrial organization by Chandler, 1977). Upon recognizing that innovation and technological change hold potential for new industries and job creation after deindustrialization, debates over the French Regulation School and flexible specialization dominated the conversation, with an emphasis on the link between various types of innovation and economic growth. Pursuing these themes further, the 1980s saw growing interest in the organizational aspects of production (with emphasis on high-technology manufacturing industries) and its geographic consequences (particularly with respect to regional and national competitiveness) (see, for example, Malecki, 1985; Scott and Storper, 1986; Castells, 1985); in addition, focus increasingly shifted towards highlighting strategies for urban and regional growth in national and international contexts (see Clark 1986; Schoenberger, 1985).

At the same time, another new branch of economic geography was emerging. This was represented by those who became increasingly steeped in work by French and German philosophers and drew significantly from political theory, critical social theory, cultural studies and archi-tecture (Dear, 1988; Harvey, 1989; Soja, 1989). This group incorporated themes from post-modernism, post-structuralism, feminist theory and cultural geography, and rejected the positivistic underlying assump-tions that had dominated economic geography to date. Furthermore,

5

the on-going globalization of economies resulted in greater interest in the socio-cultural factors that affect economic growth in various parts of the world. Some took inspiration from sociology (Castells, 1984; Giddens, 1984; Granovetter, 1985) to seek new frameworks that incorporate socio-cultural aspects of economic change. These trends combined together became known as the 'cultural turn' in economic geography.

Thus, economic geography became even more heterodox since the 1970s, with a variety of ideological orientations, such as Ricardian, Keynesian, Marxian, Polanyian, Gramscian, Schumpeterian, as well as broadly neo-classical, evolutionary, institutional, cultural and critical approaches to co-exist in the subdiscipline. Whereas area studies and ethnography have been closely associated with anthropology, political science and sociology, industry studies are conducted by scholars active in business/management and in economic sociology; and regional science is dominated by neo-classical economists. This disciplinary and ideological mix has contributed to further diversify methodological approaches, which now include econometrics and statistical analysis, questionnaire surveys, structured and semi-structured interviews, and participant observation. As economic geographers we conduct structural-institutional analysis, cultural analysis, policy discourses, social network analysis, as well as archival and textual analysis (see Barnes et al., 2007). Also, the advent of corporate geography in the 1980s meant that access to firms became increasingly important to researchers, and with data becoming increasingly proprietary, economic geographers have opted for unconventional and mixed-methods approaches in their research. Thus, methodological pallets for economic geographers have greatly expanded during this period, and have firmly moved beyond exclusively positivist economic analysis.

Economic geography since the 1990s

Economic geography since the 1990s reflected these diverse shifts that had occurred since the crises of the 1970s. The rebirth of economic geography, known as the 'new economic geography' (NEG), was claimed by different groups, including those who catered to the epistemological segment of the 'cultural turn' (e.g. post-modernism/post-structuralism) and those with econometric interests, strongly influenced by the work of Krugman (1991). The latter group of geographical economists gradually gained legitimacy for the use of the term NEG in the 2000s. At the same time, to better understand the emergence of multinational corporations (MNCs) and spatial implications, and to treat on-going globalization as an explicitly territorial process, economic geographers have attempted to link geographically specific endowments with global networks. This effort has in part been supplemented by continuing influences from economic sociology, this time in the form of global commodity chains (Gereffi, 1994).

Another recent change in economic geography has to do with geographic orientation, which has shifted from predominantly industrialized economies to include emerging economies, and there is an on-going dialogue between economic geographers and development geographers on various themes of common interests. This is a significant change for economic geography, which has been concerned primarily with building theories and gathering empirics in advanced industrialized economies. Thus, although the history of the subdiscipline has long been strongly centred around debates in and about North America and parts of Western Europe (in particular, the United Kingdom), the Anglo-American tradition in the discipline is gradually being transformed by the globalization of the subdiscipline itself. The previous dominance of Anglo-American economic geographers will doubtless give way as the next generation brings new ideas based on their far more diverse experiences.

7

As broad as the field of economic geography has been, significant new conversations and linkages are being forged today, such as that between geographical economics and economic geography, with the former mainly populated by economists and the latter by geographers.[2] Geographical economics is strictly a branch of economics, aimed at formalizing methodology with roots in international trade theory (e.g. see Combes et al., 2008). Most economists who attempt to model growth and trade across more than one spatial unit of analysis fall into this category. Analysis on the role of proximity is often optional, and the associated rise of agglomeration externalities is typically resolved by assumptions on the mobility of labour and capital. Krugman's (1991) definition of economic geography as the study 'of the location of factors of production in space' (483) exemplifies the approaches taken in geographical economics.

In contrast, economic geography remains a distinctive subdiscipline in geography that focuses on economic differences, distinctiveness and disparity across places; the political, cultural, social and historical dimensions of industrial and regional development; inter-scalar (e.g. global-to-local) economic relationships and their significance for firms, industries and regions; and the causes and consequences of unevenness in the world economy. Regardless of multiple trajectories, one of the most important and distinguishing features of economic geography from economics has been its empirical orientation. Globalization has raised new and important questions about the interconnections and interdependencies among economies, about the processes through which regions develop and compete, and about the utility of classical economic theories for explaining contemporary economic trends and challenges. Moreover, like globalization itself, a new generation of scholars is expanding the field geographically beyond Western Europe and North America in order to understand these relationships, processes and challenges.

The goal and organization of the book

This book covers the breadth of economic geography from its origins to the present and from developed to developing countries and regions, but it does so without losing sight of what has been the 'core' of economic geography. The writing of this book rose out of our collective concerns that contemporary debates have at times lost sight of what has been learned in the past; neglect of the discipline's history and conceptual roots not only leads to reinventing the wheel, it can prevent further theoretical advances in a field of study. The historical continuity and evolution we highlight in this book are intended to help students and scholars see the links between contemporary themes and earlier, often classical, work, thereby enabling the contextualization of emerging concepts in the various intellectual traditions that have characterized the field.

8

Our goal is to reconnect old ideas and new phenomena, and we do so by systematically demonstrating the relevance of historical concepts to contemporary debates and showing that what are viewed as newly emerging themes have long intellectual traditions. Each chapter offers origins, initial and subsequent conceptualizations and their transformation. In addition, we examine key contemporary problems and how each concept is used and applied today in order to provide explicit links between contemporary problems with theoretical debates. In doing so, we provide an overview of the key concerns of economic geographers, how these concerns have changed over time, how they are inter-linked and, in some cases, how they have co-evolved.

We have chosen 23 key concepts as points of departure for various ideas that are central to understanding economic geography. These concepts were chosen by their importance over an extended history of the discipline and continue to hold contemporary importance. Some key concepts are agents and drivers, while others are contextual conditions, and yet others are outcomes and spatial manifestations. Furthermore, although the prominence of each concept has waxed and waned, each provides an entry point to a variety of research questions. Collectively, these key concepts combine methodological and ideological orientations, and together they span the field.

There are many ways to organize this book. As we shall show, these key concepts have varying inter-disciplinary links. Some are closely associated with a branch of economics (neo-classical, Marxist, evolutionary and institutional economics), while others are associated with sociology, cultural studies and critical theory. In addition, some of the grand theorizing conducted in economic geography has been based on evidence from just a few industries. For example, the automobile industry has dominated the theory of Fordism/Post-Fordism whereas the garment industry has traditionally been central to the analysis of global value chains.

9

Section 1 begins with the key concepts that function as important agents of economic change: Labour, Firm and State. The relative importance of these agents has waxed and waned over time. Labour has long been a key factor input to production, forming the foundation of the drivers in Section 2. Firms emerged as a central focus particularly in the 1970s and 1980s, with influences from behavioural geography as well as from business/management studies. Firms are also an important unit of analysis for the key concepts in Section 3: Industries and Regions in Economic Change. Finally, economic geographers have long recognized the State as a key agent of economic growth, as reflected in debates on core–periphery and subsequent studies of industrial restructuring in the core along with the emergence of newly industrializing economies of East Asia.

Section 2 covers key drivers of economic change – Innovation, Entrepreneurship and Accessibility. In Innovation, we discuss three branches of economics (neo-classical, Schumpeterian, evolutionary) that have been influential in economic geography. Studies of Innovation and Entrepreneurship suggest the importance of the Schumpeterian tradition in contemporary economic geography, and they represent an important interface to economic development policy. The role of innovation and entrepreneurship is carried forward by key concepts such as Industrial Clusters (Section 3) and Knowledge Economy (Section 6). Accessibility is perhaps the most fundamental socio-geographic concept that affects mobility, and has enduring effect on the day-to-day livelihoods of people. Accessibility has also acquired new meaning with the advent of the Internet.

In Section 3 we turn to four key concepts that are the basics for understanding the role of industries and regions in economic change. The chapter on Industrial Location covers the work of many who are considered the founding fathers of the discipline. Regional Disparity is one of the oldest and recurring policy concerns, and a central motivational issue for economic geography as a discipline. Understanding Industrial Clusters is popular among contemporary scholars and policy makers, but we show that the concept has deep historical roots. In Post-Fordism, we examine the historical trajectories of dominant industrial organization (primarily based on empirical evidence from the automobile industry) and key terminologies associated with them, including Taylorism, Fordism, and Post-Fordism.

Section 4 shifts focus to key concepts that are integral in analyzing Global Economic Geographies. What are the different ways that economic geographers understand inequality in the global economy? Core–Periphery primarily covers the tradition of world-system theory arising out of structural Marxism, whereas the chapter on Globalization emphasizes the operations of multinational corporations (MNCs) in building industrial organizations at the global scale and encompasses a broad array of theoretical and empirical debates over the geographic and strategic roles of multinational corporations as well as critiques of globalisation. With Circuits of Capital, we return to a Marxian interpretation of urban and regional development with flows of capital. Finally, Global Value Chains includes the work of scholars who employ commodity chain analysis to understand how MNCs spatially organize their production and sourcing activities and whether lower-tier suppliers, who are often based in developing countries, can upgrade the value of their contributions to the global economy.

Section 5 includes key concepts that represent analytical tools to understand socio-cultural contexts. These contexts not only serve as key geographic determinants of economic activities; they also represent many contemporary views actively adopted by economic geographers today. In Culture, we review the renewed importance of culture prompted by the inter-disciplinary 'cultural turn', as well as the characteristics and critiques of post-modernism and post-structuralism. Also discussed are the subsequent development of culturally oriented research, including research on global convergence, conventions and norms, and the cultural economy. Under Gender, we address the contributions of feminist economic geography in better understanding variations in access to economic opportunity and in the economic processes that create regional diversity. Institutions help explain economic growth today, and in this chapter we include both economic and sociological interpretations of institutions. In Embeddedness our emphasis is on the analysis of social relations, illustrating how the social pervades the economic, with implications for the economic growth potential of places. Lastly, the socio-cultural contexts structuring economies can be viewed as compilations of Networks which are horizontal, flexible and infused with power relations.

The final section focuses on emerging concepts that represent economic trends of increasing significance in the twenty-first century. Although the idea of a Knowledge Economy was conceived in the 1960s, this concept has been steadily gaining significance, as the basis for newly emerging sectors of economic activities has shifted from natural to knowledge-based resources. Similarly, the Financialization of economies did not emerge overnight, but has been in progress since World War II and will gain more importance with the financial crisis of the 2008. Consumption is a relatively new focus for economic geographers, who have long been preoccupied with production. Cross-disciplinary fertilization and learning from other disciplines are therefore prominent in research on consumption today. Finally, Sustainable Development emerged in the 1980s, and we expect to see more research at the intersection of climate change and economic change in cities, suburbs and in developing countries. While these concepts are still evolving, we offer some exploratory discussions on the emerging significance of the debates that surround these concepts.

The book was truly a collective endeavour. We had numerous retreats and extensively commented on, critiqued and edited each other's writing. After the outline for each chapter was collectively agreed upon, each chapter was initially drafted by a single author and was developed further with extensive comments and edits involving all authors. Nonetheless, we

11

note the chapters' primary authors as follows: Aoyama was responsible for drafting the Introduction, Firm, Innovation, Industrial Location, Industrial Clusters, Regional Disparity, Post-Fordism, Culture, Knowledge Economy, Financialization and Consumption; Murphy was responsible for drafting State, Core–Periphery, Globalization, Circuits of Capital, Global Value Chains, Institutions, Embeddedness, Networks and Sustainable Development; and Hanson was responsible for drafting Labour, Entrepreneurship, Accessibility and Gender.

NOTES

1 The authors of the first issue included an official of the U.S. Forest Service and World War I veteran, an agricultural economist at Clark University who also worked with the U.S. Department of Agriculture, an official from Canada's Interior Department, an independent scholar and a faculty member from the University of Stockholm, Sweden.

2 For example, the new journal *Journal of Economic Geography*, in its inception (2000), was intended as a means to strengthen linkage and exchange ideas between economists and economic geographers.

Section 1
Key Agents in Economic Geography

What are the key agents of change in territorial economic transformations? Section 1 begins with the key concepts that function as important agents of economic change: Labour, Firm and State. Although the relative importance of these agents has waxed and waned over time, together they constitute vital agents that shape the reality of, as well as disciplinary interests in, economic geography.

Labour has long been a key factor input to production as well as a depository of knowledge, forming the foundation of the drivers in Section 2. Furthermore, labour has been understood as a class, and as a key unit of analysis in household reproduction, industrial bargaining, social movements and political interest groups. The locations and mobility of labour have had significant impacts on where industries emerged, and the division of labour and segmentations within the labour market have had significant impacts on variations in household incomes, productivity and social processes.

Firms emerged as a central focus in economic geography particularly starting in the 1970s and 1980s, with influences from behavioural geography as well as from business/management studies. Firms are key agents in creating jobs, commercializing innovation, and delivering new products and services to consumers. Territorial development and the location decisions of firms are in symbiotic relationship, as firms access not only raw materials, but also pools of labour with particular knowledge and skills, which in turn are shaped by the location decisions of firms. In addition, corporate strategies typically have a geographic dimension. Specializations in certain products/services or diversification to related industries both require a certain quality and quantity of labour, as well as access to a certain quality and quantity of markets. Competition induces responses of different kinds, such as product differentiation (through investment in innovation), offshoring and outsourcing, all of which have geographic consequences. Firms are therefore an important unit of analysis for the key concepts in Section 3: Industries and Regions in Economic Change, and some view the rise of multinational corporations (MNCs) as key agents of globalization (Section 4).

Finally, economic geographers have long recognized the state as a key agent of economic growth, as reflected in debates on core–periphery and subsequent studies of industrial restructuring in the core along with the emergence of newly industrializing economies of East Asia. States control access to capital (e.g. through interest rates) as well as labour (e.g. through education, immigration laws, provision of housing), which in turn influence firms' location decisions. States also manage international trade and foreign direct investment in order to ensure that jobs are created and innovation is captured within a state's territory. There is considerable variation in how states intervene in the economy, however, and striving towards an ideal form of governance has long been central to the debates in economic philosophy and economic geography.

14

1.1 LABOUR

Labour is one of the agents of economic change. Labour has been conceptualized as a key factor of production, a socio-economic class (e.g. the proletariat, the creative class), as an important agent of social movements (i.e. collective bargaining through labour unions), and a source of innovation and technological change (entrepreneurship). Although technological change has traditionally been viewed as an aspect of capital productivity, today it is widely recognized that various process innovation can significantly boost labour productivity.[1]

The mobility of labour is considered to be far more limited than that of capital, making it difficult to transfer knowledge embodied in labour from one place to another. Furthermore, although research focuses on labour performed for wages in firms, other forms of labour, often unpaid, are central to informal economies, self-employment and household reproduction. As the face of waged work and the faces of paid workers have changed over the past forty years, so too have the questions that economic geographers have posed about labour and its role in economic growth. In general, views have shifted from seeing workers as an undifferentiated input into the production process to seeing workers as heterogeneous and active agents shaping economic geographies.

Classical views of labour under capitalism

Economists have long theorized labour's role in the economy. The long-standing and most prevalent view has focused on labour as a key input to the production process, along with land and capital. Classical economists such as Adam Smith and David Ricardo considered the role of labour in creating the *value* of a commodity, and this conceptualization is called the labour theory of value.

Smith (1776), Ricardo (1817) and Marx (1867) all recognized that a commodity has both exchange-value (exchangeability to other commodities) and use-value (in which utility is realized through consumption). Smith also viewed labour as human capital, an essentially fixed factor of production and a source of knowledge. Ricardo (1817) sought to introduce more precision to Smith's assumptions and claimed that the value of a commodity is measured by the relative quantity of labour inputs. Ricardo's primary contribution was the labour theory of value, which distinguishes between the value of a commodity (exchange-value) and wages (determined by the subsistence needs of labourers).

Marx differed from Ricardo and argued that the value of a commodity is constituted as socially necessary labour-time, which is determined by average skill levels in a given state of technology in a given society. According to Marx, labour is the only source of material wealth and, as such, Marx saw the value of a commodity not as the outcome of an economic law but as a socially and historically constructed phenomenon that differs by civilization. Marx believed that labour power, or the capacity of labour, had benefitted the labour class itself before the advent of capitalism, but became commodified to be sold to the capitalist class after the means of production (e.g. land and machinery) were taken away from farmers in a violent process of primitive accumulation.[2] Marx further argued that surplus value arises when labour yields commodities whose value exceeds the wages paid; capitalism is structured in such a way that, for survival, the labour class (proletariat) has no choice but to transfer surplus value to the capitalist class (bourgeoisie), thereby producing the condition of exploitation of the proletariat. Marx also pointed out that capitalism produces structural unemployment and referred to such surplus labour as the 'industrial reserve army' (see Peet, 1975 for a detailed explanation).

16

While Marx has been criticized, for example for his focus on male industrial workers and neglect of women's role in the economy (e.g. McDowell, 1991), his main contribution has been to highlight the vulnerability of those who lack control over the means of production. Following this logic, today scholars and activists alike are concerned with the persistent presence of the 'working poor', i.e. those who work but are not earning what is considered as a living wage.

Political economy of labour and economic geography

Economic geographers have been influenced by Marxian political economy in understanding labour; most importantly, Harvey (1982) brought Marxian political economy into economic geography debates. He coined the term spatial fix to refer to a geographic strategy used by capital to resolve capitalism's contradictions, namely, the tendency for economic crisis and the falling rate of profit (see 4.3 Circuits of Capital); examples of spatial fix include the territorial expansion of production and engaging in international trade. Walker and Storper (1981) used the Marxian premise of uneven development to understand industrial location not as a static allocation of resources, but as part of the dynamics of accumulation.

Marxist approaches have shared with traditional or neo-classical ones a 'capital-centric' concept of workers, in which workers are seen from the vantage point of how capital makes location decisions (Herod, 1997). In this view labour is differentiated primarily on the grounds of varying skills, and therefore varying costs to employers. Capitalists survey the spatial variation in skill levels in the labour landscape at scales from the intra-urban to the global, and locate firms and establishments so as to gain access to the desired type(s) of labour (Dicken, 1971; Storper and Walker 1983). This view, in which labour's varying skills and costs have the ability to pull investment to certain locations, contrasts with the industrial location theory of Alfred Weber (1929), in which transportation costs were far more powerful than labour in firms' location decisions (see 3.1 Industrial Location); but as transportation costs waned relative to labour costs during the twentieth century, labour's influence on location decisions increased.

17

Doreen Massey (1984) was important in highlighting capital's role in creating spatial divisions of labour at the national scale, but Massey argued that regional disparity is the result of social processes, not solely economic ones, and that economic geography is concerned with 'the reproduction over space of social relations' (p. 16). Massey conceptualized the spatial division of labour in terms of the social relations of production (p. 67). As a basis for understanding regional transformation, the spatial division of labour involves the economics of labour markets and of social tradition, defining the power of the skilled versus the unskilled. Workers' skills, as well as regional inequalities, are socially constructed in workplace and community and are neither given nor solely the result of economic processes. Industries' and firms' location decisions respond to geographical unevenness in the labour landscape and incorporate spatial inequality in order to maximize profits; their decisions, in turn, affect workers' future skill levels and shape the future of regional economies. Structures of ownership, managerial hierarchies, the sectoral orientation of regional industries, regional cultures, gender relations within households and labour relations all play a role in this process of regional change (see 3.3 Regional Disparity). In emphasizing the social relations of production, Massey's analysis highlights the role of labour in regional transformation.

Labour under globalization

The nature of paid labour in the industrialized world has changed over the past four decades as economies in these countries have undergone significant restructuring, including shifts away from large-scale assembly-line manufacturing to smaller-batch production techniques and away from manufacturing altogether towards services (see 3.4 Post-Fordism; 6.1 Knowledge Economy). In the industrialized world during the Fordist era, working-class men in often-unionized manufacturing jobs earned a 'family wage', sufficient to support wife and children. The family wage has become increasingly rare as manufacturing jobs have relocated to emerging economies and union membership has fallen. The technical and social divisions of labour have grown ever more complex, as new groups of people (e.g. young women in Korea (Cho, 1985) or Mexico (Christopherson, 1983)) have been drawn into the labour force of multinational corporations (MNCs) (see 4.2 Globalization). The demand for low-wage labour in the service sector has increased, and many such jobs are occupied by women who have entered the labour market in mounting numbers since the 1970s in large part to shore up household incomes. It should be noted, however, that many households in the Fordist era did not have anyone earning a 'family wage' and that many women, especially women of colour, have long worked for wages inside and outside the home, and did so well before the 1970s (Nakano-Glenn, 1985; McDowell, 1991).

Capital is the factor input with mobility and agency, eschewing or even moving away from locations with high levels of organized labour (such as central cities in the 1960s and 1970s) in favour of places where labour unions are weak or non-existent (such as suburban areas) (Storper and Walker, 1983). The place-specific skills of labour affect not only firms' location decisions but also the labour process itself, that is, how work is designed within firms. For example, firms will establish electronics assembly plants entailing repetitive tasks on an assembly-line in places where workers are perceived as unskilled and undemanding, which explains the location of *maquiladoras* along the US–Mexican border to target a young, female, Mexican labour force (Christopherson, 1983). But corporate functions that demand highly sophisticated labour (such as scientists and engineers in research and development facilities) still largely remain in the industrialized world, and this is particularly evident in the case of electronics and pharmaceuticals (Kuemmerle, 1999); many developing countries still struggle to achieve the necessary skill upgrading to attract the foreign direct investment that can offer high-wage jobs (Vind, 2008).

Today, it is widely recognized that labour is no longer a homogeneous class; in fact, it has become highly diverse, with well-educated workers in professional, technical and managerial jobs earning relatively high incomes in occupations with career ladders while workers with less formal education – or international migrants who lack needed language skills or job certification (e.g. in the health professions) – earn relatively little for their labour in low-end service jobs with poor prospects for advancement (McDowell, 1991; Peck, 2001). At the same time, employers of highly skilled 'knowledge workers', in fields from the creative professions to information technology, increasingly value the ideas of such workers, realizing that they are the key to innovation and regional prosperity (Malecki and Moriset, 2008) (see 2.1 Innovation; 6.1 Knowledge Economy). The mobility of these highly skilled workers therefore influences the economic fates of cities and regions. Florida (2002), for example, found strong associations between urban growth and the creative class (see 6.1 Knowledge Economy), arguing that the geography of talent increasingly dictates the location of economic development. Saxenian (2006) discussed the role of globally mobile highly skilled labour contributing to the growth of Silicon Valley (United States), Taiwan and India.

Although overall work time has declined and leisure time has increased in the industrialized world, Peck and Theodore (2001) contend that there has been a flexiblization of labour, with temporary and contingent jobs replacing full-time, year-round employment. Indeed, what was traditionally understood as a dual labour market, comprising the primary labour market (with job security, mostly white-collar jobs) and the secondary labour market (with less job security, mostly blue-collar jobs) has broken down. These developments also render the simplistic division between capitalist and labour classes obsolete as units of analysis. Instead, labour is increasingly viewed multi-dimensionally, through the lens of ethnicity, immigration status, gender, innovation and global social movements.

The role of the state in providing support for labour has also been redefined, with many governments severely limiting welfare provisions such as workforce training programs and unemployment insurance (see 1.3 State). The state has long been influential in shaping the social relations of production and creating spatial divisions of labour. As pointed out by the Regulation School, a variety of state institutions affects the relationship between capital and labour (Peck, 1996) (see 3.4 Post-Fordism). These institutions take the form of laws and regulations prohibiting employer discrimination on the basis of a worker's sex, age or race as well as laws governing, for example, the minimum wage, welfare and workfare, hours of work and overtime pay, the right to collective bargaining, and worker health and safety on the job. Laws and practices concerning provision of child care, elder care, nursery schools and kindergartens affect the form of labour-market participation, if any, of people with care-giving responsibilities. Another way that the state affects relations between capital and labour is through education; a state's desire to develop or maintain a highly skilled labour force to drive economic development motivates public investment in education (see 5.3 Institutions). Through all of these spatially varying practices, the state helps to create the spatial divisions of labour that capital (in the capital-centric view of labour) exploits in its investment decisions.

20

Organized labour as an agent of social change

Labour is not only an active agent of economic change through unions and collective bargaining, but it has been a key driver of social change. Economic geographers have actively investigated how labour's collective activities influence economic processes, and have examined the challenges posed to union organizing by the service economy and globalization. Herod (1997, 2002) describes how the dockers' union on the US East Coast successfully fought to replace a system of local bargaining with one that covered 34 ports from Maine to Texas, thereby protecting longshoremen's jobs from the consequences of containerization. The processes of globalization are also the focus of strategies adopted by the

Canadian Auto Workers Union in the face of NAFTA and the increasing fragmentation of unions in North America (Holmes, 2004). Rutherford and Gertler (2002) observed that although union workers in the automobile industry re-scaled their bargaining approaches in response to just-in-time production, Canadian and German labour unions employed divergent strategies with varying degree of involvement in corporate decision making. Similarly, Wills (1998) suggested that globalization does not necessarily threaten labour organizations, but instead offers new opportunities for labour internationalism.

Most workers, however, work in service-sector jobs, not the traditional industries represented by longshoremen and auto workers. Walsh (2000) and Savage (2006) review the challenges of organizing service-sector workers, who are diverse in education, occupations, race/ethnicity and gender as well as being spatially dispersed throughout an urban area rather than concentrated in the large-scale workplaces of the Fordist era. Glassman (2004) sought to incorporate a Gramscian view of hegemony to better understand how the changing relationship between capital and labour resulting from globalization has affected labour unions in the United States. These authors raise questions about the geographic scale(s) at which union organizing decisions should be made as they attempt to balance a need to be sensitive to local conditions with a need to build effective trans-local institutional structures capable of challenging capital on a stage larger than the local. Work on unions demonstrates how unions are institutions that structure capital–labour relationships differently in different places (see 5.3 Institutions).

21

As Christopherson and Lillie (2005) illustrate, global labour standards are increasingly set by MNCs rather than through national regulations. But as they set such standards, MNCs are increasingly subject to the actions of consumer campaigns aimed at enforcing corporate social responsibility (Hamilton, 2009). Labour groups are therefore being supported and strengthened in their efforts by consumer activism, which has aided in the setting of workplace health and safety standards and has targeted child labour (Hughes et al., 2008; Riisgaard, 2009). Campaigns for ethical consumption challenge corporations in the networks of global value chains to rethink their labour practices, underscoring the links between production and consumption and between productive and reproductive work (see 6.3 Consumption; 4.4 Global Value Chains).

KEY POINTS

- Labour has long been conceptualized as key to creating value in the economy. Classical and Marxian economists differed, however, in the way they conceptualized labour under capitalism.
- The nature of work and the face of the labour force has changed significantly under globalization, and economic geographers' concepts of labour have become increasingly nuanced, taking into account various skill levels, new work patterns (i.e. flexiblization) as well as gender. The state's role in shaping labour has also been transformed.
- Labour's agency has been particularly apparent in the collective action of labour unions. However, the nature of collective bargaining is changing as a result of globalization.

FURTHER READING

Savage and Wills (2004) introduced a special issue of *Geoforum*, which explores the changing roles and strategies of trade unionism in an era of globalization. Boschma et al. (2009) examined the performance of plants in Sweden and analysed the role of labour mobility and skill levels.

22

NOTES

1 Labour output is typically measured by wage per worker per hour, which translates into labour productivity. Combined with capital productivity, it constitutes a portion of total factor productivity, which is often used as an indicator of, or proxy for, economic growth. Higher productivity may arise from exploitation and/or from technological change.

2 Adam Smith viewed primitive accumulation as a largely peaceful process based on individual incentives and hard work.

1.2 FIRM

In economic geography, firms have been central units of analysis in studies of industrial location, innovation, agglomeration, industry networks and regional development (see 3.1 Industrial Location; 2.1 Innovation; 3.2 Industrial Clusters; 5.5 Networks). At the individual/micro-level, economic geographers have studied business start-ups (see 2.2 Entrepreneurship), and at the global scale, economic geographers have studied the rise of multinational corporations (MNCs) and their role in innovation and knowledge transfer, in national and regional economic growth, and in the process of globalization (see 4.2 Globalization; 4.1 Core–Periphery). Firms are key agents in that they are drivers of the economic system, and their locations, strategies (including innovative capacity) and cultures have emerged as a particularly active focus of study in the past few decades.

What is a firm?

Firms are typically a collection of individuals who are organized in a hierarchy to pursue for-profit activities. Firms are started by one or more entrepreneurs (known as 'start-ups') and range from single-establishment, self-employed individuals or partnerships to multi-establishment entities with various levels of liabilities attributed to the owners. Firms can also be classified by types of ownership: state-owned enterprises, publicly traded (shareholder-owned) private enterprises for which stocks are listed and traded in a stock market(s), and privately held enterprises, for which stocks are not traded, but in which, founding entrepreneurs and high-level executives typically hold equity. State-owned enterprises played a significant role in many states particularly in the early stages of their industrialization (e.g. France, Germany, Japan, South Korea, China, Vietnam), although in advanced capitalism, private enterprises are largely responsible for sustaining the necessary dynamism in the economy.

Firm-centred studies have become particularly active since the rise of corporate geography in the 1970s (see Dicken, 1971, 1976). Large firms have significant capital and productive capacity (e.g. automobile assembly), and can dictate the market by establishing oligopolies or monopolies. Under perfectly competitive markets, firms cannot set prices, but once monopoly/oligopoly is achieved, firms can set prices, giving them opportunities to enjoy profits. Large firms, particularly those that form 'company towns'[1], where the local economies are closely associated with the dominant firm's profit cycles (see Markusen, 1985), can contribute significantly to the rise and fall of cities and regions. Conglomerates are typically giant firms that engage in multi-sectoral activities (e.g. General Electric in electronics and finance). Firms can also be networked through corporate merger and acquisitions and cross-ownership of equity (e.g. Japanese *keiretsu*, Korean *chaebols*).

The overwhelming majority of firms are categorized as small businesses.[2] Small firms are considered important to the health of the economy as they inject competition into the market to counteract monopoly/oligopoly. Because small firms are also seen as important sources of job creation and innovation, many governments offer support and incentives for small firms and new start-ups. Often small firms work in coordination with other firms in a hierarchical organization (i.e. as subcontractors to large firms) or in horizontal networks where specialized firms coordinate production. For example, in the case of industrial districts in Italy, active bottom-up coordination among small firms can serve as an alternative to mass production, with the latter being typically coordinated by large firms in a top-down manner (Piore and Sabel, 1984) (see 3.4 Post-Fordism).

With a growth of inter-firm alliances, the boundaries between firms are becoming increasingly blurred in some cases (Dicken and Malmberg, 2001) (see 5.5 Networks). For example, firms may acquire or merge with other firms and internalize functions, leading to vertical or horizontal integration.[3] Alternatively firms may opt to outsource one or more previously internalized functions to an outside firm in order to focus on core competence, leading to horizontal or vertical disintegration.[4] Firms' boundaries are further blurred by emerging forms of work organization. For example, taking the case of advertising firms, Grabher (2002) argued that business practices are increasingly organized around projects ('project ecologies'), which undercut the notion of the firm being a coherent and unitary economic actor.

Why do firms exist?

Most neo-classical economists understand the firm as an entity that takes advantage of internal economies of scale as well as internal economies of scope.[5] Economies of scale refer to the per unit cost reduction in inputs as the volume of output increases. Economies of scope refer to the cost advantages of carrying out related activities by using the same capital and labour already deployed in a factory.[6] Harrison (1992), for example, observed that flexible specialization was a means to achieve economies of scope without having to sacrifice the benefits arising out of economies of scale (see Post-Fordism).

Coase (1937), a Nobel laureate economist, explained that firms exist because they reduce 'transaction costs' or costs in using the price systems of the market, which, in neo-classical economics, are one of the criteria for imperfect competition (see 5.3 Institutions). In essence, firms emerge because of the presence of transaction costs; internalizing transactions is therefore advantageous until the cost of internal transactions equals or exceeds the cost of external transactions. Thus, according to Coase's institutional theory of the firm, firms are an alternative means of coordinating production to market-based transactions.

25

Firms and decision making

Most early studies assumed that firms operate as 'an economic man' according to the principles of rational choice. Firms are also assumed to operate by the principle of maximizing revenues and minimizing costs, and are always in search of achieving greater profitability. Yet, economic geographers have long known that discrepancies exist between the reality and the rational choice based on the principle of economic man. For example, Törnqvist (1968), Watts (1980) and Dicken and Lloyd (1980) all examined biases in locational decisions that were inexplicable by classical industrial location theories (see 3.1 Industrial Location).

The behavioural theory of the firm emerged from organizational theory in business/management literature, which also aimed at addressing the discrepancies between theory and reality. The notion of bounded rationality developed by Simon (1947) was one of the first attempts to reassess and redefine human actions via organizational behaviour. Given the overwhelming complexities of the real world, the rationality behind decisions is often limited by the knowledge and experience of the decision makers. Simon's work broadened the assumption to include not only optimizing but also satisficing principles[7] in understanding firm behaviour. The reality of decision making is better represented as rationally bounded and context-dependent. Following Simon, the work of a prominent business historian Alfred Chandler (1962, 1977) gave legitimacy to the studies of decision making and its impacts on organizational form in modern corporations. In sum, human agents are far less competent in calculations and less trust-worthy than the model of economic man implies, and bounded rationality and opportunism may better characterize the decisions made by firms.

Williamson (1981) sought to explain the presence of firms and hierarchical organizations, through illuminating the presence of bounded rationality and opportunism in arms-length market transactions. He argued that there are a number of conceptual barriers among various perspectives – neo-classical economists' lack of interest in firm hierarchy, organizational theorists' neglect of the market, policy analysts' scepticism over monopolistic/oligopolistic firms, and economic historians' neglect of organisational innovation. Extending the work of Coase, he proposed to view the firm as a governance structure that economizes transactions costs. According to Riordan and Williamson (1985), firms' 'make-or-buy decision' (i.e. whether to produce internally or purchase from external sources) for necessary parts or services is explained by asset specificity of the firm, which results in cost differentials between sustaining internal governance versus relying on market transactions. Scott (1988) also adopted the transactions cost approach and argued that firms will attempt to balance scale and scope effects with market prices of inputs and outputs, and develop an optimal internal organization accordingly.

Williamson argued that a study of firms is a study of comparative institutional undertaking, which take place under certain explicit or implicit contractual framework. This perspective diverges from the purely economic perspective and incorporates social and cultural dimension of the firm's operation. Through contracts, both explicit and implicit, firms institutionalize trust to minimize transactions costs, which explains why oligopolies and monopolies are particularly effective in reducing transaction costs.[8]

Denzau and North (1994) suggest that, instead of self-interests, 'people act in part upon the basis of myths, dogmas, ideologies, and 'half-baked' theories' (p. 3). According to them, decision making under uncertainty is effectively an outcome of a co-evolutionary process between collective institutions and individual mental models, the former being both formal and informal, and the latter characterized by a degree of 'shared subjectivity' acquired through learning. As a result, firms are known to have different corporate cultures and their own internal social dynamics (see 5.2 Gender).

Firms are also understood as strategic entities. The resource-based (also known as competence-based or knowledge-based) theory of the firm strives to understand how enterprises gain sustained competitive advantages by developing strategic resources that enable them to exploit new market opportunities (Penrose, 1959; Barney, 1991; Barney et al., 2001). A firm's human capital (technical and managerial skills), physical capital and organizational capital (less-tangible assets) serve as resources to coordinate knowledge management, customer relations, planning and production. Coordination within firms invariably involves tacit knowledge, but Maskell and Malmberg (2001) believe that such knowledge will eventually be codified and ubiquified in order for firms to take advantage of it effectively. Maskell (1999) in particular emphasized how firms structure learning processes, and act as repositories of knowledge, especially when being different from others is an important competitive foundation of the firm. Firms are therefore integral to the creation of localized capabilities for place-specific competitiveness (Maskell and Malmberg, 2001).

In economic geography, firms are increasingly viewed as socio-cultural entities that produce certain social regulations (Schoenberger, 1997; Yeung, 2000). This view goes against the traditional economic view of the firm, which sees firms exclusively as profit-maximizing entities. Schoenberger (1997) discussed the cultural crisis of the firm, drawing examples from such companies as Lockheed and Xerox, and demonstrated how corporate cultures emerge out of managerial identities and commitments, which in turn produce corporate strategies. Corporate culture becomes particularly complex as firms establish offices across borders and evolve into MNCs. In some cases, the cultural attributes of the firm's origin (entrepreneur) and the firm's headquarter location have a significant impact on corporate culture, whereas in other cases, the national origin of the firm is increasingly viewed as irrelevant once it becomes transnational.

27

The rise of multinational corporations (MNCs)

MNCs refer to corporations that own and operate production and/or service functions in more than one country. The East India Company, a trading firm established by the Dutch is among the first MNCs. Trading companies dominated MNCs until the post-World War II period, when foreign direct investment (FDI) began growing at a rate faster than international trade. Unlike trading companies that specialized in the trade of mostly raw materials to be processed in the developed world, new MNCs conducted manufacturing across borders, with commodity chains crossing various international borders, i.e. New International Division of Labour (see 4.2 Globalization). Active offshoring of productive facilities, primarily to low-wage locations, began in the 1950s and 1960s and continues today. While this process still forms the core conceptual rationale for the existence of MNCs, contemporary firms are not only seeking low-wage locations; they are also seeking access to certain skills or access to new markets (i.e. Toyota and BMW in the United States; Microsoft and Intel in China). These market-seeking MNCs often form joint-ventures with local firms to gain local market knowledge and knowledge of relevant government regulations.

28

Why do MNCs exist? The transactions cost explanation posits that MNCs exist because the vertical and horizontal integration of various functions results in cost advantages. The same paradigm sees the MNC as an effective governance structure, minimizing transactions costs by internalizing functions that are otherwise particularly costly when conducted across borders. Alternatively, Dunning (1977) proposed an 'eclectic theory' that seeks to explain MNCs' locational behaviour through the OLI framework, which includes Ownership (or firm-specific), Location and Internalization advantages. Dunning argued that combinations of these advantages determine whether FDI will take place, where it will go to and how it will enter the market.

In the late-1980s and early-1990s, policy debates on whether to support entry of MNCs for job creation ('free trade') or to protect national competitiveness by offering incentives to domestic over foreign MNCs ('managed trade') (see Reich, 1990; Tyson, 1991) led to various studies on the role of nationality of MNCs in their strategic goals (Encarnation and Mason, 1994). Phelps (2000) argued that increasingly innovative institutional responses are needed for local host economies to attract

MNCs. Recent research on MNCs includes locational analysis of specific functions such as corporate headquarters (Holloway and Wheeler, 1991) and research and development (R&D) facilities (Florida and Kenney, 1994; Angel and Savage, 1996) as well as the role of MNCs in latecomer industrialization, with specific focus on emerging Asian MNCs (see Yeung, 1997, 1999; Aoyama, 2000; Lee, 2003)

Instead of acquiring more capacity through the acquisition of plants overseas, firms today are increasingly opting for outsourcing, i.e. forming contractual relationships with firms on a project-by-project, or product-by-product basis, rather than internalizing productive capacities through greenfield investment or through corporate merger and acquisition. While firms can outsource to another domestic firm, outsourcing today is typically arranged with firms in low-wage locations often in another country (see 4.4 Global Value Chains). Outsourcing emerged as a particularly important strategy for fashion-oriented clothing and shoe brands as a way to increase agility and flexibility, reduce financial liabilities and at times protect reputation by distancing the brand from potentially damaging practices of outsourced contractors. However, outsourcing has also become controversial as it is increasingly viewed as a strategy to offload corporate social responsibilities to foreign contractors. In order to reduce child labour and sweatshops, anti-sweatshop movements increasingly insist that not only the fashion brands but also their contractors abide by certain labour and environmental standards (Donaghu and Barff, 1990) (see 6.3 Consumption). Research on corporate social responsibility movements have expanded to cover MNCs in agri-food business and retailing (Hughes et al., 2007).

29

KEY POINTS

- Firms have long been central to research in economic geography. Recently, however, new views on the firm are emerging, such as networks and cultural/institutional views of the firm.
- Primary assumptions on the behaviour of firms have also been transformed. Bounded rationality, opportunism and cultural factors that influence decisions of the firm have been explored.
- Spatial strategies of firms under globalization, and in particular, locational decisions of MNCs, have been analysed extensively. Most recently, outsourcing practices and locational strategies of MNCs are emerging as an important area of study.

FURTHER READING

Maskell (1999) offers a comprehensive overview of theories of the firm. For a recent work on firms, see Fields (2006) for a study of Dell Computers, and Monk (2008) for how performance of US automakers was linked to legacy costs. Beugelsdijk (2007) examined the link between firm performance and regional economic environment in the Netherlands.

NOTES

1 Company towns refer to cities and towns whose economies are largely dependent on one or a few dominant firms.

2 The definitions of small businesses (also known as small and medium enterprises – SMEs) vary across countries, but commonly firms with less than 500 employees are considered small. According to the U.S. Census, there were 25,409,525 firms in the United States in 2004, out of which only 17,047 firms had more than 500 employees. This means 99.93 per cent of firms fell under the small business category.

30

3 When the new function is in the same commodity/value chain, the integration is vertical, whereas horizontal integration occurs when the newly acquired functions belong to another product/commodity chain.

4 In the early 1990s, scholars in business/management began arguing for firms to focus on their 'core competence', i.e. identifying the bundle of competitive strengths and creating new businesses. Prahalad (1993), for example, observed that the core competence of Sony, a Japanese consumer electronics giant, was miniaturization.

5 External economies of scale refers to benefits of scale experienced collectively among a number of firms, typically in the same locations (see Industrial Clusters).

6 For example, one firm that produces an audio-stereo system may also produce a portable CD player, without significant additional cost in adding new equipment or training workers.

7 Under satisficing principles people seek adequate, not optimum, solutions.

8 Various studies on trust emerged in the 1990s. For example, see a discussion by Williamson (1993) on calculative trust; Sako (1992) on the role of trust in arms-length and contractual relationships; and Sabel (1993) on trust as negotiated loyalty and studied consensus.

1.3 STATE

States play central roles in structuring economic and industrial development, and they exist in a diverse array of place-specific political ideologies, economic institutions and state–society relationships. There are two broad perspectives on the state in economic geography. The first views it as an extra-economic force that intervenes to correct market failures and guide economic processes. The second views the state as being inseparable from capitalist systems and the geographical context where it is situated. Economic geographers have studied how different varieties of capitalism have evolved historically in particular places, how inter-state trade and investment agreements shape the global economy, and how the territorial strategies of states influence the distribution of economic opportunity and power within them.

States-in-capitalism or capitalist-states?

Economic geographers have developed important ideas and frameworks for understanding how states relate to or influence capitalist development. Glassman and Samatar (1997) describe three distinct perspectives. Liberal-pluralist perspectives view states as autonomous from economic institutions and thus able to serve as 'neutral arbiters' over questions of resource distribution, the provision of infrastructure and other classical state functions (e.g. military decisions, social services). Marxist perspectives disavow the notion that states are neutral, seeing them instead as agents and institutions whose actions are principally aimed at sustaining capitalist accumulation through the continuation of unequal exchange and the exploitation of labour power. Neo-Weberian perspectives take a middle position, viewing the state apparatus as deeply connected to or embedded in socio-economic institutions but able to act autonomously to address non-economic issues such as welfare needs of workers, families and communities (see 5.4 Embeddedness).

Clark and Dear (1984) made an important distinction between theories that emphasize the role of states in capitalism versus those concerned with how, where and why particular forms of capitalist states evolve. The states-in-capitalism approach focuses on how the actions of states influence economic activities. In this literature, the role of states is traditionally confined to the provision of public goods (e.g. education, military defence) and the strategic intervention in capitalist processes in order to correct or prevent market failures; i.e. circumstances where negative social, environmental and/or economic consequences or externalities result from 'natural' market-driven activities (Arndt, 1988).[1] Key debates in the states-in-capitalism literature deal with how much state intervention is appropriate (Lall, 1994; Manger, 2008), how the economic policies of states often create or exacerbate environmental problems and socio-economic inequalities (Hindery, 2004), and what kinds of state intervention can best foster employment, regional growth and redistribution (Haughwout, 1999).

For many geographers, however, research on capitalist states and their constitution is of greater interest. In this view, capitalism is seen as a historically contingent and politically contested form of socio-economic organization and the priorities and institutions that constitute contemporary states reflect historical class struggles and context-specific factors such as culture, geography and religious ideology (Jessop, 1990; Dicken, 2007). States and capitalist systems are viewed to co-evolve, thus making it impossible to fully understand markets or governments in isolation from one another. Recent studies on capitalist states have analysed how they have modified their institutions and economic policies in response to the demands of neo-liberal forms of economic globalization (Jessop, 1994; Glassman, 1999; see 3.4 Post-Fordism; 4.2 Globalization).

Varieties of capitalism

The capitalist state literature demonstrates how capitalism is not a monolithic or coherent system but a diverse set of ideologies, institutions and forms of economic organization that vary geographically. Interest in the different forms of capitalist state inspired the varieties-of-capitalism literature, which is concerned with the embedded nature of the states, and how state actions are distinctly socio-cultural and reflect particular market ideologies (e.g. see Albert, 1993; Berger and Dore, 1996; Hall and Soskice, 2001). For geographers, the varieties-of-capitalism idea offers an important conceptual lens for explaining differences among states, their historical and spatial evolutions, and how citizens and ties to other states shape such evolutions (e.g. see Das, 1998). Four varieties of capitalism generally describe most states: welfare states, developmental states, socialist states and neo-liberal states.

Keynesian welfare states or social democracies manage market forces by mitigating the effects of cyclical economic downturns, maximizing employment levels and sustaining the tax revenues needed to provide public goods and services such as social security, education, health insurance and national defence (Brohman, 1996). Welfare states use monetary and fiscal policies to sustain economic growth and to redistribute wealth across society. In some cases, states may partially own critical industries such as banking, steel and automobile manufacturing in order to protect jobs and ensure that backward (to raw material suppliers) and forward (to value-added processors) linkages remain within the country's boundaries.

Developmental states, commonly associated with East Asian economic powers such as Japan and South Korea, aggressively intervene in domestic industries in order to pursue export markets and tightly control import flows. While close ties between economic and political elites are highly problematic in most countries, successful developmental states achieve close business–government relationships principally through large industrial groups characterized by cross-ownership of equity (*keiretsu* in Japan, and *chaebol* in South Korea). Developmental states specify 'target industries', which are typically key industries with extensive backward and forward linkages in the domestic economy (Johnson, 1982). For example, South Korea's investment in steel production was closely followed by the development of shipbuilding and automobile manufacturing (Amsden, 1989). Evans (1995) explains the success of some developmental states such as South Korea and India as an outcome of embedded autonomy, a situation where the state and the private sector agree on how industrial development should proceed but the state remains autonomous enough to define national economic objectives while domestic firms remain efficient and competitive. Developmental states do provide basic services (e.g. health, education) within their borders, but the explicit priorities of rapid modernization and high growth often come at the expense of social and spatial inequalities and democratic rights (e.g. the Three Gorges Dam in China; South Korea's history of dictatorship from the 1960s to the 1980s; gender discrimination in Japan). 33

Communist and socialist states are intensively controlled in order to centralize political power, reduce social class distinctions and create self-sufficient or autarkic economies that are delinked from capitalist trade relations. Socialist states are typically single-party states which tightly control property and means of production. Socialism is understood as a necessary stage in the move toward communism, in which the state would eventually relinquish its power in favour of communal control. In this sense, communism has never been successfully achieved at the national scale; as a result, most so-called 'communist' countries are better described as socialist economies. Centralized economic planning (in which prices are determined by the state, not by the market) is common in most socialist states along with, in some cases, industrial clustering (e.g. see Lonsdale (1965) on the Soviet system of territorial production complexes). In addition, welfare distribution and spatial equality receive considerable attention as stated goals although achieving these goals has proven elusive. Few purely socialist economies survive today (e.g. China, Cuba), and since the collapse of the Soviet Union many formerly socialist economies have begun the transition toward a market economy. These transitional economies (e.g. Poland, Vietnam) face unique challenges as they strive to develop efficient market systems while maintaining some of the entitlements and social services that were guaranteed under socialism.

Since the 1980s, neo-liberal states have become common in the world economy. In this case, the state remains significant for defence and some welfare programmes but it takes on a *laissez-faire* (i.e. non-interventionist) stance towards social and economic issues. Jessop (1994) refers to the emergence of neo-liberal states as the 'hollowing out' of the welfare state, with strategies entailing fiscal decentralization, privatization and a shift from social welfare (guaranteed entitlements) to workfare programmes (you get what you work for) (Peck, 2001). Much like the developmental state, neo-liberal states promote export-oriented industrialization (EOI), which seeks to exploit a country's comparative advantage in international markets (World Bank, 1994). The shift to EOI strategies is a key distinction between neo-liberal and Keynesian welfare states, given that welfare states often rely on import-substitution industrialization (ISI), which aims to protect domestic industries through trade tariffs and quotas on imported goods.[2] In promoting EOI, neo-liberal states work in conjunction with international financial institutions (IFIs) such as the International Monetary Fund (IMF) to increase their access to foreign markets, to deregulate global capital flows and to guarantee the property rights of their companies when they invest in foreign countries. EOI strategies have been adopted by developmental states, and in most developing countries they have been promoted through the structural adjustment programs imposed by multi-lateral donors (e.g. the World Bank). Structural adjustment programmes strive to transform domestic financial institutions, markets and regulatory systems to promote free trade. Since the neo-liberal model has become widely popularized, social inequality has increased significantly in many developing and advanced industrial economies, as well as globally (Arrighi, 2002; Gilbert, 2007).

Interstate relations and trade and investment agreements

In recent years a shift can be observed away from politically driven interstate alliances (e.g. NATO) to those driven by economic objectives and priorities. Specifically, states are increasingly engaging in bilateral, multi-lateral, and regional trade and investment agreements that enable them to manage international capital and resource flows more effectively. These agreements generally take on one of three forms: double-taxation treaties (DTTs) that prevent individuals or firms in two countries from being doubly taxed on dividends or profits; bilateral investment treaties (BITs) that strive to improve investment flows and the security of investments made between two countries; and preferential trade and investment agreements (PTIAs) that strive to achieve greater economic coordination and market access for participating countries. In some cases PTIAs may involve more than economic issues as evidenced by the US government's Andean Trade Promotion and Drug Eradication Act (ATPDEA) that provides preferential trade opportunities for Andean countries such as Bolivia, Colombia and Peru provided they actively reduce coca leaf cultivation and circulation.

35

Regional economic unions and trade agreements such as the European Union (EU) and the Agreement of Southeast Asian Nations (ASEAN) are also playing an increasingly important role in the governance of the global economy. These agreements typically link geographically proximate countries to one another with varying levels of economic integration and political coordination (Dicken, 2007). Free trade areas (e.g. the North American Free Trade Agreement (NAFTA)) eliminate trade barriers between participating countries, with little or no political or economic coordination beyond what is needed to reduce trade and investment barriers. Customs unions (e.g. Andean common market (CAN)) enable free trade, and participating countries also pursue a common external trade policy with non-members. Common market (e.g. the Caribbean Community (CARICOM) or East African Community (EAC)) are areas where all factors of production, such as labour and capital, are able to move freely between members. Finally, there is the most advanced form of regional economic integration, the economic union (e.g. the European Union). Unions require a high degree of economic and political coordination; member-states may be expected to support a common financial system, a common currency, and the establishment of common trade, investment, environmental and other regulations.

By the end of 2007, there were 2,608 active BITs, 2,730 DTTs and 254 PTIAs. As the circles and ellipses in Figure 1.3.1 demonstrate, most countries and regions rely on a complex blend of bilateral, intra- and inter-regional PTIAs when engaging with the world economy and some have described the situation as a global 'spaghetti bowl' of political-economic alliances (UNCTAD, 2005). Importantly, these complex relationships and interdependencies belie the WTO's rhetoric about the possibility for an equal, free-market-driven global trading system. Most of all, they demonstrate how states continue to play a significant role in the world economy.

Territoriality: states in action

Regardless of their form or variety, all states operate at a variety of scales (e.g. the urban, regional or national) as they set and enforce regulations, determine and guarantee property and other civil rights, provide for and manage public goods (e.g. education, defence, the environment), mobilize society, and strive to ensure that labour power is reproduced and that its productivity increases over time (O'Neill, 1997). In doing so, states as territorial entities strive to increase their power and legitimacy in their territories as well as relative to other states (Cox, 2002). The territorial strategies of states are implemented formally (e.g. in labour laws, environmental regulations or taxation policies) and informally (e.g. through corrupt practices, nepotism), and they create the socio-economic conditions that influence industrial and regional development processes.

36

Because achieving consensus on territorial strategies is often a struggle, states are best understood as social arenas where competing actors put forth alternative economic visions or imaginaries. Such struggles create both 'winning' and 'losing' ideas, firms, industries, workers and communities (O'Neill, 1997; Peck, 2001). Strong actors mobilize power both within the state's territory and at other spatial scales (Jones, 2001; Brenner, 2004; see 4.2 Globalization; 4.3 Circuits of Capital). Local growth coalitions, national industrial promotion agencies, public–private partnerships, international development organizations and chambers of commerce – groups that link private sector actors and state officials – often play a key role in driving policy changes or in preventing shifts away from the status quo (Cox, 2002).

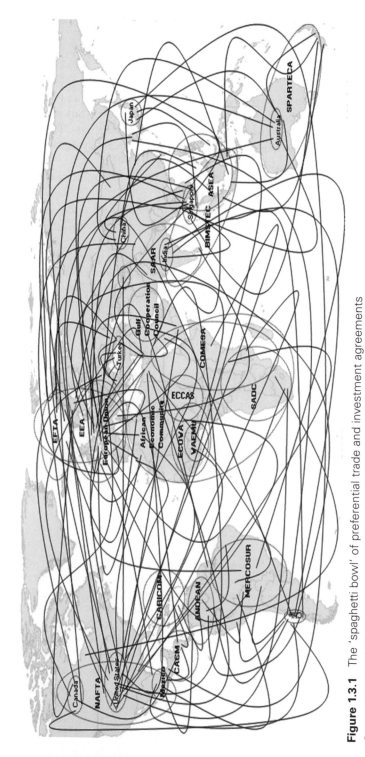

Figure 1.3.1 The 'spaghetti bowl' of preferential trade and investment agreements

Source: UNCTAD, 2005, http://www.unctad.org/en/docs/iteiite200510_en.pdf © United Nations, 2005. Reproduced with permission.

KEY POINTS

- There are two broad perspectives on the state in economic geography. The first views the state as an extra-economic force that intervenes to correct market failures and guide economic processes. The second views the state as being inseparable from capitalist systems and the geographical context where it is situated.
- The governance strategies of states reflect different varieties of capitalism and state–society relations.
- Bilateral and regional trade and investment agreements are being increasingly used by states to gain competitive advantages in the global economy.
- States act territorially to create socio-economic conditions and to organize economic activities that will ideally increase their power and legitimacy within and beyond their formal boundaries.

FURTHER READING

Peck and Theodore (2007) provide a detailed review and sympathetic critique of the varieties-of-capitalism literature. Harvey (2005) reviews the history of neo-liberal forms of capitalism and provides an assessment of its failings and of the possibilities for alternative forms of state–society relationships to evolve in the coming years. Hollander (2005) details the territorial strategies of sugar growers in Florida who appealed to national-security concerns in the federal government in order to ensure that trade barriers and subsidies would protect the sugar industry.

NOTES

1 Three types of market failures are inherent to capitalism: the tendency for competition to become monopolistic over time, the free-rider problems associated with the management and regulation of public goods (e.g. defence, the environment) and negative market-driven externalities such as pollution or labour discrimination.

2 A central goal of ISI policies in developing or emerging economies has been to nurture infant industries able to sustain employment while enabling domestic firms to develop the capabilities needed to be more competitive in global markets.

38

Section 2
Key Drivers of Economic Change

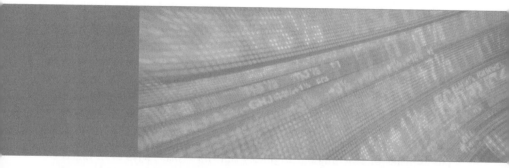

What are the sources of dynamism in an economy? Section 2 covers key drivers of economic change – Innovation, Entrepreneurship and Accessibility. In Innovation, we discuss three branches of economics (neo-classical, Schumpeterian and evolutionary) that have been influential in economic geography. As society has become increasingly dependent on technology, innovation has become ever more indispensable. Yet, innovation extends well beyond the realm of technology to shape economic consequences at various scales, including the individual (entrepreneurship), the firm, the region and the nation (the national innovation system) as well as the broader global economy. Innovation has come to be viewed as a key ingredient in economic growth, initiating the formation of new economic sectors, creating new opportunities for entrepreneurship and contributing to job creation.

Although everyone recognizes the importance of entrepreneurship in injecting dynamism in the economy, entrepreneurship is also difficult to pin down conceptually. This difficulty has led to a challenge in translating conceptual knowledge into practice, in the form of entrepreneurship promotion. As we shall show, entrepreneurship serves multiple purposes and exhibits a variety of characteristics. Economic geographers agree, however, that entrepreneurship is deeply connected to place, and therefore contexts, culture, social networks and economic systems (such as taxation or unemployment benefits) all affect the emergence of an entrepreneur. Studies of innovation and entrepreneurship suggest the importance of the Schumpeterian tradition in contemporary economic geography, and they represent an important interface to economic development policy. The role of innovation and entrepreneurship is carried forward by key concepts such as Industrial Clusters and Knowledge Economy.

Accessibility is perhaps the most fundamental geographic concept that affects the economy; it has an enduring effect on the day-to-day livelihoods of people. While physical distance may define people's daily access to workplace or school, shops and housing, transportation technology and associated infrastructure have dramatically expanded individuals'geographic reach. The way that economic geographers have conceptualized accessibility has changed over time, from simple distance to one that overcomes not just geographic, but also economic and social barriers. Accessibility has also acquired new meaning with the advent of the Internet. Whereas virtual space offers unprecedented possibilities in reaching out to places that have otherwise been inaccessible, new barriers to accessibility, such as the digital divide, are also evident.

2.1 INNOVATION

Innovation involves successful commercialization of an invention, a new idea or technology (a bundle of knowledge), the eventual marketing of which contributes to the development of new economic activities and regional economic growth (see Freeman, 1974; Dosi, 1982). The definition of innovation has become broader over the years. Instead of considering as innovations only technological breakthroughs (such as the invention of electric light-bulbs), various forms of innovation are recognized today, including product innovation, process innovation (new ways of making existing products, such as just-in-time production of automobiles), design innovation (new designs/packaging of existing technologies), incremental innovation (as opposed to breakthroughs), and service innovation (e.g. new ways of selling existing products, such as fast-food restaurants). Distinctions are also made between a new-to-the world innovation vs. a new to the area innovation. Although the latter represents the application of an existing idea in a new geographic context and has been understood as a technology transfer rather than an innovation, it is a form of innovation diffusion and can generate significant local impacts (Blake and Hanson, 2005). Geographers initially focused on analysing the spatial diffusion of innovations (see Hägerstrand, 1967; Rogers, 1962; Brown, 1981), but more recent research has shifted to innovation in the context of regional and national economic growth; this strand of work has been heavily influenced by branches of economics that consider technology as an engine of economic growth.

It is difficult to count or measure innovation. The most frequently used measures of innovation include research and development (R&D) spending (public and private), the number of approved patents and patent citation data, and the size of the science and technology labour force. However, some of these measures only serve as proxies, while others only take into account successful outcomes (innovation) and ignore failed attempts.

Technology in neo-classical economics

To neo-classical economists, technology has long been an unexplained aspect of the production function.[1] For example, the Harrod-Domar model (Domar, 1946; Harrod, 1948) suggested that productivity growth in an economy comes from capital accumulation (e.g. purchase of more machinery) combined with the level of savings. Solow (1957) incorporated technological change in his growth model and explained that what had been regarded as the 'residual' is in fact responsible for productivity growth in the long run. He observed that capital accumulation alone cannot explain the sustained productivity growth of the United States between 1909 and 1949. Solow, however, regarded technological change as an exogenous factor.

Among the modifications proposed within the neo-classical economics framework, the best known is Arrow's work that incorporated knowledge as an endogenous variable in an economy's production function. Arrow argued that 'learning' – the acquisition of knowledge – is the product of experience that 'can only take place through the attempt to solve a problem' (Arrow, 1962: 155). From this, he concluded that technical change emerges out of experience in production, which in turn raises overall productivity. Learning-by-doing is therefore a process by which accumulated experience in production leads to productivity gains for a firm.

42 Arrow's conceptualization of learning has led to the distinction between tacit and codified knowledge. Whereas codified knowledge is explicit, tacit knowledge is implicit, context-dependent and can only be shared through socialization. Michael Polanyi (1967), brother of Karl Polanyi, was among the first to elaborate on tacit knowledge, who conceived it as 'know-how' learned through experience, which is often difficult to articulate. Due to the experiential nature of tacit knowledge, codified knowledge cannot completely replace it.

Lundvall and Johnson (1994) developed the notion learning-by-interacting to emphasize the role of tacit knowledge in innovation. They contend that the production of tacit knowledge is simultaneously the very act of its transmission. As a result, geography plays a key role in the production of knowledge critical for innovation, because search is typically conducted locally first. Whereas the business/management literature highlights the role of the firm in the production of tacit knowledge, economic geographers are interested in broader socio-economic contexts that shape the rules of action. The importance of geography is reinforced through the mutual sharing of trust, norms and conventions, which ultimately develop into 'communities of practice' (see 5.4 Embeddedness). In addition, following Lundvall (1988), Gertler

(2003) points out that the exchange and production of tacit knowledge also take place not only among producers, but also between users and producers.

The Austrian school of economics

Quite separate from other neo-classical economists of the time, Joseph Schumpeter (1939) was developing a set of distinct long-run views of economic growth through his work on business cycles. He saw capitalism as being full of discontinuous change (Schumpeter, 1928), a view that is captured in his hallmark concept of 'creative destruction', which he considered essential to capitalism (Schumpeter, 1943). For Schumpeter, not every invention qualified as an innovation; instead, a true innovation must have a significant impact on productivity through fixed capital investments and should lead to the rise of new firms. Entrepreneurs are risk-taking actors endogenous to an economy, who are distinct from capitalists and inject instability into capitalism. These fast-moving, quick-response entrepreneurial small firms create new leadership in the industry. Entrepreneurs seek to capture short-run monopoly gains from innovation. This leads to imitation behaviour among entrepreneurs and causes *swarming*,[2] which creates new industries and amplifies economic growth (see 2.2 Entrepreneurship). Schumpeter (1942) was concerned that monopolistic tendencies would lead to innovation to take place increasingly in the R&D facilities of large firms, which would produce 'routinized' innovation and certain 'technological regimes' that would hinder the necessary process of creative destruction.

43

The revival of Schumpeterian economics, prompted by the world economic slowdown of the 1970s, also revived some business-cycle theorists of the bygone era. Among the notable business-cycle theorists revived by neo-Schumpeterians were Kondratieff and Kuznets, both of whom attempted to predict the cyclical nature of economic growth and decline over the course of history. Unlike Kuznets cycles, which were 14–18 years, long and were measured by the patterns of national investment (see Kuznets, 1940), Kondratieff (1926, 1935) hypothesized that economic booms are induced by major waves of technological innovation, which cycle through an economy at intervals of 54–57 years. Theorists predict that the fifth Kondratieff 'wave' (by those who did not like the rigidity implied by the term 'cycle'), one that is based on the electronic revolution, will end in the 2010s (Hall and Preston, 1988).

Evolutionary economics

Nelson and Winter (1974) sought to build on Solow's work on technological change and economic growth and incorporated ideas offered by Schumpeter and ultimately developed what is known today as Evolutionary Economics. The goal was to develop a dynamic equilibrium model based on a behavioural theory of the firm (see 1.2 Firm). In this model, decisions of micro-agents (firms) are assumed to be based on bounded rationality and are therefore heterogeneous but also exhibit behavioural continuity of decisions ('routines'). The inner workings of technological change is understood as a dual process of 1) firms engaging in local *search* for better technologies, and 2) the markets engaging in the *selection* process. This search and selection process is combined with satisficing principles to generate a path of technical change.

44

Arthur (1989) and David (1985) argued, through the notion of path-dependence in technical change, that markets do not always select the best or the most efficient technology. Using the case of typewriter keyboard sequence QWERTY, which was deliberately designed to slow down typists in order to avoid jamming, David (1985) showed that norms, habits and practices defined by and designed for technologies of the past can live on long after initial technological constraints have been removed and even if the solution is highly inefficient. In the case of the keyboard sequence, path-dependence is reinforced by 'technical interrelatedness' among habits, practices and institutions that support and promote the practices, such as educational institutions that offer training for typists. Through this case study David demonstrates how the quasi-irreversibility of investment, economies of scale, and chance and historic accidents can shape technological trajectories.

Standard setting has become an important competitive strategy for firms because once a standard is set, the cost of switching technologies becomes increasingly prohibitive for users, thereby reinforcing an already existing technological trajectory. Arthur (1989) called the process lock-in. Arthur emphasized two properties in his dynamic approach;

non-ergodicity which suggests that historical chance events may decide the outcome (a dominant technology), and inflexibility which refers to the process that an outcome becomes progressively more locked-in over time. Notable examples include competing computer-operating systems (Microsoft Windows versus Apple Macintosh), or VCR formats (the mid-1970s war between two types of video tapes).

For evolutionary theorists, the question of whether successful innovation takes place is therefore dependent upon a variety of factors, including opportunities (sources of technical knowledge), presence of incentives, capabilities of firms and mechanisms for implementation (Dosi, 1997). Evolutionary economics has been used to support industry life cycle theory (Utterback and Abernathy, 1975) to illustrate historical co-evolution of technology and industry structure and to explain how dominant design through economic natural selection destroys variety and reinforces mechanisms for path dependence. Evolutionary economics contributed to the emergence of the techno-economic paradigm in the field of business/management (Dosi, 1982). In this paradigm, technology is defined as a knowledge set that includes know-how, methods, procedures, physical equipment, and experiences of successes and failures, and can be both practical solutions and theoretical applications. A technological trajectory is viewed as the outcome of multi-dimensional trade-offs among various technological options, and is shaped by economic (the firm's search for new profit opportunities), cultural and institutional factors (industrial organization, regulations). Freeman (1988, 1995), for example, developed the concept of national systems of innovation, in which technological trajectories are examined along with the role of formal and informal institutions in the coordination and promotion of technology transfer. He contrasted features of the national systems of innovation in 1970s Japan and the former Soviet Union and argued that, in spite of globalization, national institutions continue to play a fundamental role in innovation.

45

Evolutionary economic geography

Evolutionary economic geography is an attempt to develop a spatialized application of concepts developed in evolutionary economics. Boschma and Lambooy (1999) were among the first in articulating how evolutionary thinking contributes to understanding the process of localized collective learning, the decline of regions, and the emergence of new industries in particular places. The field is still relatively new, however, and its approach and methodology are evolving. In fact, Essletzbichler and Rigby (2007) claim that although various concepts of evolutionary economics, such as path-dependence, routine, lock-in and co-evolution, have now been widely adopted in economic geography, as a theoretical framework evolutionary economic geography still lacks specificity.

Although primarily concerned with the interaction between techno-logical change and economic geography, there are nuances in how scholars interpret the main focus of evolutionary economic geography. According to Boschma and Martin (2007), the primary concern of evolutionary economic geography is on how processes and mechanisms of adaptation and novelty make or hinder the geographies of production, distribution and consumption, and vice versa. Martin and Sunley (2006) focus on how micro-behaviours of economic agents culminate in self-organization, and how the process of path creation and dependence interact with the geographies of economic development. Others (Freeman, 1991; Essletzbichler and Rigby, 2007) bring back the Darwinian dimension of evolutionary theory by claiming a long history of interaction between biology and social sciences; these authors focus on the role of various selection environments, including the natural and the institutional. Esseletzbichler and Rigby (2007) emphasize the role of competition in shaping micro-agents' behaviour, and argue that economic agents and places co-evolve to shape distinctive geographic landscapes.

An increasing synergy can be observed between evolutionary and institutional economic geography (Essletzbichler and Rigby, 2007; MacKinnon et al., 2009). Especially with the emphasis on learning as an important factor in innovation and transfer of tacit knowledge, research on innovation in economic geography is increasingly taking on an institutional dimension. Storper (1997) argued that innovation and institutions are increasingly seen as integral aspects of a 'co-evolutionary process'. Feldman and Massard (2001) highlight the importance of the role of universities and favourable public policies in generating knowl-edge spillovers. Institutional variety and localized learning explain divergent paths in innovation and economic growth. Thus, by the turn of the millennium, evolutionary economic geography had cross-fertilized with institutional economic geography so that innovation research has come to incorporate strong institutional dimensions and institutional analysis has incorporated path-dependence (see 5.3 Institutions).

New trends in innovation research

Broadly, there are two theories for the drivers of innovation – 'demand-pull' or 'technology-push'. The former suggests that market forces are the main driver whereas the latter assumes technologies are knowledge-driven and emerge quasi-autonomously. Today, there is a tremendous interest among industries and firms to exploit the innovative capacity of the users through user-led innovation. Although research on user-producer interactions has a long history (see, for example, von Hippel, 1976; Lundvall, 1988), the Internet has fundamentally altered how user–producer interaction can bring about innovation. This is particularly the case of lead users, who as first-mover consumers, adopt products early in their product life cycle, typically have expert knowledge of the products, and play the role of trend-setters (Porter, 1990). The Internet has allowed direct peer-to-peer interactions en masse, as exemplified by open-source software development (e.g. Linux, Mozilla Firefox). Firms are motivated to involve their customers more actively, not only by the need to accurately predict and capture latent demand, but also by the potential to reduce R&D costs (von Hippel, 2005). By integrating the consumer, businesses are able to offer product diversity while minimising the risk of failure. Grabher et al. (2008), for example, proposed a typology of co-development involving customers based on empirical evidence from firms in Germany and the United States.

47

KEY POINTS

- Ideas about innovation have evolved dramatically in the last several decades. Today, the definition of innovation has been broadened significantly to include not only radical breakthroughs, but also incremental innovations. In addition, process innovation is now recognized in addition to product innovation.
- Research on the role of geography in innovation has also evolved significantly in the past few decades from a focus on diffusion to an interest in learning as a key factor endowment for economic growth. Innovation is viewed as a highly contextual activity, which is shaped not only by economic, but also by social and institutional factors. Innovation is seen as cumulative, path-dependent and context-specific.

- Evolutionary economic geography is focused on the interactions among technological change, path dependence and the geography of economic growth. Institutional variety shapes technological trajectories, at the local, regional and national levels.

FURTHER READING

For an extended and contemporary discussion on tacit knowledge, see Gertler (2003). For the most recent debate on evolutionary economic geography, see Martin (2010), as well as the round table with MacKinnon et al. (2009) and commentaries by Grabher, Hodgson, Essletzbichler and Pike et al.

NOTES

1 Production function is a term in economics that specifies the relationship between the combination of inputs (capital, labour) and outputs.
2 Swarming can be defined as a rapid emergence of a large number of actors populating the market simultaneously and heightening competition.

2.2
ENTREPRENEURSHIP

Many scholars view entrepreneurship, which entails the launching of a new business, as the engine of innovation and the key to understanding regional prosperity and economic change. Research on entrepreneurship is multi-disciplinary, with economic geographers joining scholars from business and management, economics and sociology to understand entrepreneurial processes. The broader entrepreneurship literature has increasingly recognized the importance of context, and to a far lesser extent place, in understanding how people start and expand businesses, indicating a growing role for economic geography in studies of entrepreneurship. Key themes and debates concern the definition of entrepreneurship, entrepreneurship and place, entrepreneurial networks and the social identity of entrepreneurs.

What is entrepreneurship?

There seems little doubt that entrepreneurship entails uncertainty and risk, but scholars have failed to agree on a single definition of 'entrepreneur' or 'entrepreneurship'. This definitional difficulty reflects fundamentally different views of what entrepreneurship is all about. Acs and Audretsch (2003: 6), for example, consider all businesses that are new and dynamic, regardless of their size or line of business, to be the result of entrepreneurship.[1] Others stress that the term entrepreneurship should be reserved for businesses that are innovative (although as Blake and Hanson (2005) have argued, 'innovative' is another term that is difficult to define unambiguously); in this view, firms that replicate the products, services or business models of other firms are not considered truly entrepreneurial (for a review of the difficulties involved in defining entrepreneurship, see Gartner and Shane, 1995). This latter view stems from the influential work of Schumpeter (1942), who saw entrepreneurship in the form of innovative new businesses as the key driver of the 'creative destruction' that he considered essential to capitalism (see 2.1 Innovation).

For Schumpeter (1936), the novelty introduced by entrepreneurs is the key to the entrepreneurial process. This novelty brings change, and thereby the instability that Schumpeter so valued, into any one of five different places in the value chain by creating (1) new products or services, (2) new production methods, or (3) new ways of organizing an industry, or by discovering/creating (4) new geographical markets or (5) new raw materials. Schumpeter (1942) also believed that such innovation is best carried out by small firms that can move quickly to develop an opportunity once identified; he saw the bureaucratization of large corporations as antithetical to entrepreneurship and as responsible for the rigidities that would spell the death of capitalism. In fact, Acs et al. (1999) show that the dynamism created by the churning (births and deaths) of small firms, whether innovative or not, is correlated with economic growth at the national level.

A longstanding view of entrepreneurship, which is moot on the issue of innovation, turns on the idea of recognizing and exploiting opportunity (e.g. Kirzner, 1973; Stevenson, 1999). Sarasvathy et al. (2003: 142) define an entrepreneurial opportunity as 'a set of ideas, beliefs, and actions that enable the creation of future goods and services in the absence of current markets for them'. These authors identify three dimensions of entrepreneurial opportunity, each based in a different market dynamic. In *opportunity recognition* the entrepreneur takes advantage of the market as an allocative process and exploits existing markets by matching existing supply and demand. In *opportunity discovery* the entrepreneur takes advantage of asymmetries of information (that exist in part because information is place-sticky) to identify a demand or a supply that does not exist (and that can be matched to a demand or a supply that does). In *opportunity creation* the entrepreneur creates, through interactions with others, both demand and supply. This last process sees entrepreneurship as embedded in, and endogenous to, ongoing social interactions. Sarasvathy et al. (2003) note that the usefulness of any one of these understandings of entrepreneurial opportunity will depend on particular circumstances.

Entrepreneurship and place

Where does entrepreneurship come from? As Shane and Ekhardt (2003) note, early attempts to explain the origins of entrepreneurial activity looked to the personal characteristics of entrepreneurs themselves, emphasizing, in particular, their propensity to take risks. From this perspective, which has not been notably successful in explaining the incidence of entrepreneurship, 'entrepreneurship depends on differences among people rather than differences in the information they possess' (Shane and Ekhardt, 2003: 162). Discontent with this 'traits approach' has led analysts increasingly to search for the origins of entrepreneurship within the contexts in which people make decisions about starting or running a business (e.g. Schoonhoven and Romanelli, 2001; Acs and Audretsch, 2003; Sorenson and Baum, 2003). Although 'context' is not always used in the geographical sense,[2] paying attention to the relationship between entrepreneurship and place makes sense insofar as most entrepreneurs launch their new ventures on home turf, for reasons explored below (Birley, 1985; Reynolds, 1991; Stam, 2007). A focus on context-as-place instead of personal traits also opens the potential for policy to promote entrepreneurship as a basis for regional economic growth. To be effective, such policy must be based on a firm understanding of the relationship between entrepreneurship and place.

With others, economic geographers have been particularly interested in how entrepreneurship, however defined, can drive regional growth. Because virtually all start-up firms are small, attention has focused on small and medium enterprises (SMEs), defined as firms with fewer than 500 employees. Birch (1981) was among the first to document that SMEs account for the majority of new jobs in the US, and others have since pointed out that SMEs employ more than half of the US work force (in 1994, SMEs accounted for 53 per cent of private-sector employment, 47 per cent of sales and 51 per cent of value added (Acs et al., 1999:11)). SMEs continue to constitute a high proportion of US firms (more than 99 per cent of all employer firms in 2004 had fewer than 500 employees), a proportion that is sustained through continuous entry (and exit) of firms rather than by an unchanging set of small enterprises (US Census Bureau, 2009). Many of these new firms are spin-offs from larger companies, enabling an entrepreneur to exploit previous experience in a large firm; this spin-off process contributes to the clustering of similar or complementary economic activities (see 3.2 Industrial Clusters).

In recent years, research on the relationship of entrepreneurship to place has advanced on several geographic scales. At the national scale, the Global Entrepreneurship Monitor (GEM) has collected data every

51

year since 1999 on entrepreneurship in selected countries around the world, growing from ten countries in 1999 to 40 in 2006, with an expected 56 countries in 2009. In a series of country reports and special topics reports (e.g. on gender and entrepreneurship), the GEM uses the data from their representative samples within each country to assess national levels of entrepreneurship, the impacts of entrepreneurship on national growth, and the relationship between national-level contextual variables and levels of entrepreneurship. In addition, dozens of scholarly articles have made use of the longitudinal GEM data, including several by economic geographers (e.g., Bosma and Schutjens 2007, 2009). These studies show that there are striking and persistent differences in rates of entrepreneurship across countries, owing to factors ranging from rates of return on investment or cultural views on risk-taking to national differences in the regulations that stand between a prospective entrepreneur's idea and its implementation. As of 2008, the highest early-stage entrepreneurship rate in the industrialized world was observed in the United States, followed by Iceland, South Korea and Greece (Bosma et al., 2008).

52 Economic geographers have been leaders in theorizing those aspects of places at the sub-national scale that appear to be related to new-firm formation, especially firms in innovative sectors such as high technology (Malecki 1994, 1997).[3] In an early example, Chinitz (1961) proposed that entrepreneurship would be more difficult in areas dominated by oligopolies (e.g. Pittsburgh at the time of his study) than in places where industries are organized competitively. In a more recent example, Keeble and Walker (1994) used county-level data for the UK to show that the factors influencing new-firm formation include demand-side variables such as previous growth in income and population density, as well as supply-side variables such as occupational and sectoral structures, firm size and the availability of capital. Keeble and Walker's analysis demonstrates, moreover, that firm size affects the founding of manufacturing and services firms differently: in manufacturing, the local presence of many small firms is positively related to growth in the number of new businesses, but for firms in professional and business services, it is the regional presence of large firms that matter. The availability of venture capital is another attribute of place that affects business formation and success, through the provision not only of capital but also of advice and oversight (Zook, 2004). Malecki's (1997) review of previous studies leads him to similar conclusions about the

place-factors that are important to innovative entrepreneurship: regional industrial mix, spatial concentration or agglomeration, the presence of skilled labour and networks that provide access to technology and capital.

While more difficult to measure and to analyse quantitatively, regional culture is another dimension of place that is closely tied to entrepreneurship (see 5.1 Culture). Contrasting California's Silicon Valley with Massachusetts's Route 128 area, Saxenian (1994) showed that the distinctive ensemble of business practices, norms, customs and beliefs in each place helped to explain Silicon Valley's greater entrepreneurial dynamism; in particular, she found that the 'risk culture' of Silicon Valley, which made business failure acceptable or even enviable, was key to developing the culture of innovation there.[4] Flora et al. (1997) developed the concept of 'entrepreneurial social infrastructure' (ESI) to capture the elements of place that foster entrepreneurship; they demonstrated that places with high ESI had developed a range of civic institutions (rather than government structures) that encouraged risk taking and venture creation. In a comparison of two regions in Japan, Aoyama (2009) shows how a distinctive cultural legacy in each place, especially regarding openness to outsiders, has shaped contemporary entrepreneurship in the information-technology sector. All of these examples illustrate the power of formal and informal institutions to create regional cultures that shape entrepreneurial outcomes (see 5.3 Institutions).

53

Although there is agreement that people tend to start businesses in places they are familiar with, relatively few empirical studies have focused explicitly on the relationship between entrepreneurship and place at the local scale. The findings of those that have done so highlight the importance of the individual's multi-faceted relationships with the community (Birley, 1985; Reynolds and White, 1997; Kilkenny et al., 1999; Jack and Anderson 2002; Stam 2007). The significance of community milieu to business formation is underlined in a study that examined the effects of violence on venture creation at the neighbourhood scale in five large US metropolitan areas; Greenbaum and Tita (2004) found that in previously low-crime areas, a surge of violence does discourage the births of new businesses, but violence surges in high-violence neighbourhoods have no impact on new-firm formation.

In a review of transnational entrepreneurship Yeung (2009) has critiqued the 'flat surface' ontology of many studies of environment–place relationships and has advocated instead a relational ontology that focuses on entrepreneurial relationships maintained through networks. An earlier and different critique had pointedly observed that people, not places, start businesses (Reynolds and White 1997), prompting many (e.g. Thornton 1999; Nijkamp, 2003) to advocate approaches that examine the interaction between individual-level and contextual processes in business formation. Both lines of thinking lead to a focus on networks.

Networks and entrepreneurship

Like other economic actors, entrepreneurs and potential entrepreneurs are part of networks of social and institutional relationships. These relationships can facilitate entrepreneurship because they are conduits for information and resources, and insofar as such relationships are based in trust, they can also be a means for reducing some of the risk entailed in starting and running a business. Networks of personal relationships have an interesting property that often benefits entrepreneurs; namely networks span and connect apparently disparate realms of everyday life over a wide range of spatial scales, linking, for example, coaching the neighbourhood youth soccer league, serving in a national professional association, running a high-tech manufacturing firm, and having attended a certain university. The boundary-spanning property of some networks means that examining only 'business relationships' is somewhat artificial because in many cases a relationship can be both personal and professional. In fact, in a detailed study of the networks of entrepreneurs running graphic design firms in Glasgow, Shaw (1997) found that business owners want to do business with people with whom they can have a friendship relation (see 5.5 Networks).

Nevertheless, many studies of entrepreneurial networks have focused on business relationships (e.g. ties with customers, suppliers, subcontractors, legal and accounting firms) and have tended to look at the networks of firms rather than individuals (e.g. Bengtsson and Soderholm, 2002; Schutjens and Stam, 2003; Kenney and Patton, 2005). A question that has motivated much of this work concerns the geographical reach of entrepreneurial networks in order to understand the extent to which entrepreneurship is a localized process. Schutjens and Stam (2003) studied the business relationships of young firms in the Netherlands and learned that three years after founding, most of the business relationships were located within the region, with extra-regional relationships declining relative to intra-regional ones over time. They also confirmed that, despite their focus on business relationships, one-third of the sampled firms said that the source of these relationships was social (p. 131). In another study, Kalantaridis and Zografia (2006), who examined the nature, duration, location and intensity of entrepreneurs' network ties in Cumbria, England, decided that the answer to the question of whether or not entrepreneurship is a local process depends on the definition of local. If local means proximate territory, then entrepreneurs who were native to the area (most of whom were craftsmen) had 'the closest linkages with their immediate geographic context'; if local means network proximity, then in-migrants (most of whom had professional/managerial backgrounds) were the most embedded entrepreneurs, as they maintained strong ties with people outside of Cumbria, mainly in the places where they had lived previously. These authors, as well as Hess (2004) and Yeung (2009), stress that network embeddedness can have a significant non-local component (see 5.4 Embeddedness).

Nijkamp (2003) believes that to be successful, an entrepreneur must develop and manage effective networks. Yet very little work has examined the connections between entrepreneurial networks and entrepreneurial outcomes, the exception being networks' impact on firm location. Pallares-Barbera et al. (2004) document how entrepreneurs' longstanding familial and social relationships within a rural region of Catalonia, Spain, created a 'spatial loyalty', which led people to launch businesses there and to create a new cluster of firms in food, textiles and machinery. Stam's (2007) study of new, privately owned, fast-growing manufacturing and business services firms in the Netherlands shows how entrepreneurial networks affect the location of the firm. Stam found that whereas 55 per cent of the 5–11-year-old firms in his study had moved since start-up, only 4 per cent had moved more than 50 kilometres/31 miles; 'most entrepreneurs who considered moving out of their home region decided against such a move because of highly valued personal relationships' (p. 46). As firms age, however, Stam believes that these personal relationships will become relatively less important vis-à-vis other factors in an expanding firm's location decisions. In a very different context, Murphy (2006) demonstrates how entrepreneurs in Mwanza, Tanzania's furniture sector adopt different location and networking strategies based on the market segments they

55

cater to. Specifically, low-value, high-volume producers rely on less-personal and low-trust relationships and locate their firms in highly visible and accessible areas of the city. In contrast, high-value, low-volume producers rely on trusting personal relationships with key clients and prefer more isolated or secluded locations for their production and retailing activities.

Place, networks and the social identity of the entrepreneur

A focus on entrepreneurial networks calls attention to the reality that networks are shaped in and through people's social identities. Because entrepreneurship, like the labour market more generally, is segmented along lines of gender and ethnicity, entrepreneurial networks within a place function differently for men and women and for members of different racial/ethnic groups (Blake, 2006; Hanson and Blake, 2009) (see 1.1 Labour). That is, in part because women and members of immigrant and ethnic groups tend to interact with other in-group members and to lack access to the networks of the dominant group of entrepreneurs (which, in the UK, Australia and North America is white men), their relationships to place and their patterns of business formation are different from those of white male entrepreneurs.

Within the US, for example, the proportions of a region's population in particular ethnic groups affects rates of business ownership among those groups, presumably because of the institutional structures that have developed related to ethnic/immigrant populations there (Wang and Li, 2007). Many studies, primarily by sociologists in the 1970s and 1980s, examined the spatial clustering of ethnic firms within US cities (e.g. Waldinger et al. 1990). But in her study of Chinese producer services in Los Angeles, Zhou (1998) concludes that not all of these Chinese-owned firms clustered in ethnic enclaves; she shows that the industrial sector of an enterprise as well as the ethnic identity of the owner affect firm location. In other regions of the world, the strong ethnic embeddedness of entrepreneurial networks has been shown to determine the business choices made by Indonesian entrepreneurs (Turner, 2007) and to affect entrepreneurial practices, especially those entailing mobility, among women traders in Benin (Mandel, 2004).

KEY POINTS

- Entrepreneurship is considered to be key to innovation, regional prosperity and economic change. Analysts differ in their definitions of entrepreneurship, with one view seeing all business start-ups as evidence of entrepreneurship, and the other considering only innovative or fast-growing businesses as entrepreneurship.
- Economic geographers have investigated the links between entrepreneurship and place, with particular emphasis on the place characteristics that appear to be related to entrepreneurial activity. The availability of skilled labour, regional industrial and occupational structures, especially with respect to firm size, and regional cultures have all been shown to be related to entrepreneurial outcomes.
- Entrepreneurial networks also play a key role in processes of entrepreneurship. The nature and geographical reach of such networks affect firm location and may affect entrepreneurial success. Because networks are comprised differently and function differently for different kinds of people, particular attention has been paid to the role of ethnic- and gender-specific networks in entrepreneurship.

57

FURTHER READING

Thornton and Flynn (2003) provide an overview of entrepreneurial networks and place. Wang (2009) shows how contextual factors at the metropolitan scale affect rates of self-employment for ethnic and gender groups in the US. For insights into the structure and impacts of transnational entrepreneurial networks, see Saxenian (2006); de Soto (1989) describes entrepreneurship in Peru.

NOTES

1 The definitional challenge here is to specify what is meant by 'dynamic', as virtually all new businesses are dynamic in some sense.
2 Among non-geographer scholars of entrepreneurship, 'context' is often used to refer to industrial, sectoral or temporal context – not place.
3 As Nijkamp points out, the emphasis in the entrepreneurship-and-place literature has been on the formation of new firms, to the neglect of the

spatio-temporal processes that enable firms, once launched, to survive and thrive (Nijkamp, 2003).

4 More recently, Saxenian (2006) has shown how Chinese and Indian migrants to the US have effectively diffused the Silicon Valley 'culture of entrepreneurship' by returning to their home countries to launch new enterprises there.

2.3 ACCESSIBILITY

In the most general sense, accessibility is the ease of reaching destinations. The term can be used for people to describe the ease with which they can reach places they want to go, such as hospitals, schools, shops, workplaces or national parks; accessibility can also be used in reference to places, to describe how easily one place or location, say for a business, can be reached by people in other places. Although in the early days of economic geography, access was conceptualized in terms of simple distance as one of the key concepts underlying the spatial organization of the settlement system (Christaller, 1966) and industrial location (Weber, 1929), accessibility also depends on the communication and transportation networks that facilitate interaction.

As a driver as well as an outcome of economic change, accessibility is important to the locations of firms, households and industries and is at the heart of social and economic well being; without access neither people nor places can survive, much less prosper. You only have to recall how a city-paralysing blizzard or hurricane literally brings daily life to a standstill to appreciate the fundamental importance of accessibility to economic and social life. Without accessibility, neither production nor consumption processes can take place; the concept is absolutely central to the 'geography' in economic geography.

The study of accessibility and mobility has also been central to the work of transport geographers, which often overlap with economic geography (see, for example, Leinbach and Bowen, 2004; Aoyama and Ratick, 2007). In addition, the work of economic geographers has intersected recently with interests among some sociologists on the social aspects of accessibility (for example, Cass et al., 2005). Furthermore, some observers have noted a methodological shift in accessibility research, particularly the increasing prevalence of qualitative analysis in understanding accessibility (Goetz et al., 2009). The thinking of economic geographers about access has shifted significantly over the past half century, from understanding accessibility primarily in terms of the distances separating people and places and the ability to overcome those distances via mobility to thinking about access in terms of the ability to connect with information, knowledge and people. Instead of being accepted as a given, the importance of distance to accessibility is now a focus of inquiry.

Accessibility and mobility

Traditionally, accessibility has been closely linked to mobility, but the two are not the same. Whereas the former refers to ease of reaching destinations (the ease with which someone can obtain medical services or library services, for example), the latter refers to the ability to move between places (the ability to travel to a doctor's office or library, for example). Accessibility is closely related to the spatial arrangement of activity sites, such as homes, workplaces, shops and schools, whereas mobility depends on the ability to move between these places, whether on foot or via bicycle, horse, automobile, bus, subway or aeroplane. Another way to think about mobility is that it requires energy to overcome the friction of distance separating an origin and a destination. The Internet has fundamentally changed the relationship between accessibility and mobility.

Before the advent of telecommunications, accessibility depended upon the ability to move from one place to another whether on foot or via other means of transportation. You could not obtain the dental or library services offered in a distant location without physically moving to the dentist's office or the library. Clearly land use patterns affect this kind of access. If medical and library services, along with schools, workplaces, shops and parks are all within walking distance of your home, you can enjoy good accessibility to the necessities of daily life without engaging in a great deal of mobility. If, however, different kinds of land uses are not located close together – as they are not in many US settlements where residents must travel many miles to reach the nearest food store or post office – then access requires considerable mobility, usually via motorized vehicles.

People around the world have been increasingly gaining accessibility through greater mobility. In the US, the number of passenger vehicle miles travelled (VMT) per annum has increased steadily, such that in 2005 VMT was more than 2.5 times the VMT in 1970 (Transportation Research Board, 2009). Worldwide, the same measure of mobility more than quadrupled between 1960 and 1990 and is expected to more than quadruple again by 2050 (Schafer and Victor, 2000). Because throughout much of the world, people have been gaining accessibility through motorized mobility and because such mobility depends on petroleum and contributes significantly to greenhouse gas emissions, the close relationship between accessibility and mobility continues to attract a great deal of scholarly and policy attention. One long-standing question that remains salient in urban planning and economic geography is how to provide accessibility without ever-increasing mobility. Changing the densities and the land use configurations of settlements is one approach that has attracted considerable attention (Transportation Research Board, 2009).

The use of various types of telecommunications, particularly the Internet, is another way to bring at least some degree of access without mobility to people and places who have access to those telecommunications devices. Obtaining accessibility via the Internet depends, however, on the type of good or service desired; for example, while most library services are readily available over the Internet, certain medical services, such as having a tooth pulled, still usually requires travel to a dentist's office. Cairncross (2001) and others have predicted 'the death of distance', the idea that, because of the increasing power of telecommunications, overcoming the friction of distance will soon require hardly any mobility, thereby robbing distance of its power to shape human activities. The data cited above on the relentless increase in mobility, however, do not support the proposition that distance or the desire to traverse it is in its death throes. The rapidly changing relationship between accessibility-with-mobility and accessibility-without-mobility is a topic of great current interest and is addressed in a later section of this chapter.

61

Economic geographers have become increasingly sensitive to the reality that access entails more than simply the ability to overcome distance, whether via the automobile or the Internet; barriers of language, education, culture and laws control access as well. The following sections will examine (1) approaches to measuring accessibility, (2) the role of networks in accessibility, (3) mobility and access to employment, and (4) the interplay between virtual and grounded accessibility.

Measuring accessibility

Most measures of accessibility assume that access entails mobility and focus on the time and money costs entailed in overcoming distance to reach destinations. Measures of the accessibility levels of people and of places are similar in that they both begin with the map, that is, the spatial configuration of human activity on the landscape plays a central role in each. Accessibility measures for people generally focus on one type of destination, say medical facilities, and count the number of such facilities within varying distances of an origin, usually the person's home, with more distant facilities being counted as less valuable because they are more difficult to reach. A simple example is

$$A_i = \sum_j O_j \, d_{ij}^{\,-b} \tag{1}$$

where A_i is be the accessibility of person i, O_j is the number of opportunities (in this example the number of medical facilities) at distance j from person i's home, and d_{ij} is some measure of the separation between i and j (this measure could be distance in km, travel costs in euros or travel time in minutes). The size of the negative exponent b on the distance term measures the difficulty of travel in a particular time and place; the larger the value of b, the more costly it is to traverse distance (compare walking through a rainforest, which would have a relatively high b-value with driving on an interstate highway, which would have a relatively low b-value). This basic equation can be altered, for example, to take into account the varying sizes of medical facilities, to reflect the likelihood that large ones at any given distance are more attractive because they offer a greater range of services.

62

This equation can also be used with data aggregated to areas, such as counties or census tracts, to assess the relative levels of accessibility of different places or areas. In this case, A_i is the accessibility of, say, census tract i, O_j is the number of opportunities (say medical facilities) in tract j, and d_{ij} is a measure of separation between tracts i and j. Note that this way of measuring accessibility – whether for individuals or for areas – gives one indicator of the overall accessibility of a particular location (origin) by summing up the number of opportunities located at varying distances from that origin location.

Although simple accessibility measures like the one described here have been widely used for a long time, especially to assess the access of places (Black and Conroy, 1977), they have their limitations, especially when used to assess the accessibility of individuals. Often the data on the nature of opportunities are insufficiently detailed; if, for example, the goal is to measure access to employment opportunities, the opportunities included in the equation should be only those that match the skill level of the person in question, but such detailed information may not be available. Even if such detailed data were available, along with point-specific location data and detailed distance data over a transportation network, these measures capture accessibility only from a single, specified origin; but as people move around over the course of a day, their access to opportunities shifts as they move, a fact that is particularly salient for assessing access to non-work activities such as shopping. Kwan (1999) has used a network-based GIS approach to calculate the set of all opportunities that can feasibly be reached, given someone's shifting location over the day and the amount of time that person has available for travel. Measures like this are more realistic indicators of access for individuals.

Transportation networks and access

Because movement on a network is easier than movement off a network, networks shape accessibility, and changes to a network that affect the cost of movement have consequences for the location of economic activity. Early work by Garrison and colleagues (1959) showed, for example, that the increased accessibility resulting from highway improvements in the western part of the state of Washington – improvements that reduced the cost of mobility – led to the growth of larger settlements and the demise of smaller ones in the network. Of course, changes to network links can also decrease accessibility, as in the case of traffic congestion, the loss of a bridge or tunnel to a flood or terrorist attack, or the cessation of airline service in a city. The point is that accessibility, and hence economic growth or decline, is closely tied to network conditions and to the location of a person or place within a given network.

63

The overall accessibility of any particular node (e.g. city or town) in a network is a good predictor of the amount of economic activity there. For this reason decisions about the configuration of the railroads, the interstate highway system, the airline hub-and-spoke system and the Internet are hotly contested because they have enormous consequences for the economic prosperity of places.

As networks have been extended and improved (by, for example, replacing a two-lane road with a four-lane highway), people can travel longer distances in a given amount of time than they could before; this process has the effect of bringing places closer together, an outcome known as time–space convergence (Janelle, 2004). At the global scale, the falling cost of movement has fuelled increases in international trade and underwritten the success of China's export industries, for example. As petroleum costs increase, the economic geography of manufacturing is likely to shift, with production sites for products that are costly to transport relative to their value, such as furniture and steel, moving closer to markets (following the classical location theory of Alfred Weber (Weber, 1929) (see 3.1 Industrial Location).

While the physical accessibility provided via transport networks is likely to remain a central consideration in the location of manufacturing activity, the importance of spatial access to the location of other types of economic activity, especially services, is less certain as the ease of interaction via telecommunications networks grows. Note, however, the sustained importance of network connectivity for any type of interaction, including virtual exchanges.

Access and inequality: the daily journey to work

64

Economic geographers' longstanding concern with the relationship of inequality to spatial processes (see Introduction) explains the large body of research on the relationships among residential location, employment location and access to jobs via the daily journey to work. Many studies, conducted at many different times and in many different locations, have documented systematic variations in the willingness or the ability of different types of workers to travel various times or distances to work. Income, employment status (full time or part time) and gender have all been shown to affect the length of the work trip, with higher-income and full-time workers and men travelling more time and/or distance than lower-income, part-time workers and women (just a few examples of these studies are: Madden, 1981; Hanson and Johnston, 1985; Blumen and Kellerman, 1990; Crane, 2007). Marital status affects women's and men's work trips differently, with marriage essentially shortening women's trips and lengthening men's (e.g. Song Lee and McDonald, 2003; Crane, 2007); in a similar vein, elder and child care responsibilities (usually borne by women but sometimes too by men) shorten the journey to work in some places, as Song Lee and McDonald (2003) found for Seoul, Korea.

In all of these studies, the length of the journey to work is taken as an indicator of the area within which a worker searches for a job and therefore of access to employment opportunities; a smaller area, indicated by a shorter work trip, suggests the possibility that a job searcher is able to access relatively few jobs. However, a major shortcoming of these studies, which rely on large secondary data sets such as those collected by a census, is the lack of information on whether a short work trip in any given instance is the result of choice or constraint. Lacking such information, interpretation of short vs. long work trips is difficult, and investigating this question of choice or constraint should be a high priority for economic geographers.

The question of how to interpret a long work trip (i.e. as an indicator of access to a wide range of employment options or a burden indicating poor access) is especially salient in assessing the impact of race and ethnicity on access to employment. An early study by economist John Kain (1968) identified a spatial mismatch between the location of manufacturing establishments, which at that time in the US had begun suburbanizing, and the location of manufacturing workers, many of whom were African American men living in the central city. Race-based discrimination in suburban housing markets prevented these workers from moving to the suburbs, and the public transportation system, which was geared to bringing suburban workers to the central city, did not serve the mobility needs of those many inner-city unemployed workers who lacked access to a car. Those inner-city workers who did travel to the suburban manufacturing jobs that matched their skill levels had exceptionally long commutes. Kain's work called attention to the importance of residential location and the (lack of) residential mobility in limiting job accessibility for racialized minorities.

65

Since Kain's influential work, several economic geographers have analysed race and ethnicity (in addition to gender) in the context of access to employment for workers with relatively fixed residential locations. McLafferty and Preston (1991) found that Latina and African American women service workers in New York City commuted the same distances as their male counterparts and significantly farther than white women did; these results contrast with studies of mainly white women and men, which document that women's work trips are shorter than men's (e.g. Hanson and Johnston, 1985; Crane 2007). Johnston-Anumonwo (1997) found that, in comparison with other race/gender groups, African American women in Buffalo, NY had both shorter and longer commutes, all to low-wage jobs; those with long commutes travelled from the central city to the suburbs, enduring a long commute for very low wages. In such cases, where people commute long distances to low-wage jobs, it is hard to see a long work trip as an indicator of good employment access.

In her study of native-born black and immigrant women in Los Angeles, Parks (2004a) confirmed that having better spatial accessibility to jobs is associated with less unemployment for native-born black and some immigrant groups. Parks (2004b) also examined the impact of access to employment on the labour-market segregation of immigrants into ethnic niche jobs, that is, jobs held primarily by co-ethnics; she found that living in an ethnic enclave is associated with working in an ethnic-niche job, and this is especially the case for immigrant men and women who live closer to immigrant-niche jobs. Immigrant women, however, were much more likely than immigrant men to work in ethnic-niche industries (that is, they experience much higher levels of labour market segregation), a finding she links to women's higher propensity to find employment through ethnic networks. For other studies on the role of networks in connecting people with jobs and thereby in helping to create and sustain labour market segmentation, see Hanson and Pratt (1991), Hiebert (1999) and Wright and Ellis (2000) (see Networks).

66

These studies of workers' mobility demonstrate how workers' decisions about residential location (which, for many workers are highly constrained) and the journey to work (which is also constrained by time, money, and the transport modes and facilities available) affect their access to employment. Can access to the Internet provide greater equality of access to employment?

Interplay between virtual and grounded accessibility

Will the Internet and other forms of telecommunications truly eradicate the friction of distance and render physical mobility obsolete? Will access via the Internet replace access via mobility? The advent of high-speed information technology (IT) has changed the organization of many economic activities by enabling coordination across long distances and thereby increased efficiency. The internationalization of some economic activities – such as call centres, back offices and computer programming – has prompted speculation that physical accessibility, which requires some degree of spatial proximity, is losing (or, by some accounts, has already lost) its power to affect the location of economic activity and to shape economic change (Malecki and Moriset, 2008).[1]

But others observe that physical accessibility retains fundamental importance, even in the age of the internet; in other words, the death of distance has been greatly exaggerated. For example, predictions that telecommuting (working from home via the Internet) would eliminate a sizable proportion of all daily work trips have not yet been borne out by the data. In fact, most studies show that use of IT tends to be complementary to, rather than a substitute for, physical mobility, and in some cases it actually increases rather than decreases travel (Mokhtarian and Meenakshisundaram, 1999; Mokhtarian, 2003; Janelle, 2004). Indeed when IT is substituted for mobility (using the Internet instead of going to the library; emailing a business partner instead of having a meeting), it can relax some of the space–time constraints on an individual's mobility (Kwan and Weber, 2003). But, as Kwan and Weber point out, for any person, access to IT varies by time of day and day of week, and time spent using IT is time that cannot be spent on other activities.

Additional insight on the interplay between virtual and grounded accessibility comes from detailed studies of various modes of interpersonal interactions within and between firms. In a recent study of the financial services industry in London, the authors found that face-to-face contact remains the preferred form of interaction, with telecommunications such as email and video being used to supplement, not replace, fact-to-face contacts (Cook et al., 2007). For workers in this type of knowledge-intensive industry, mobility is still required to access jobs. 67

But in some business transactions, once trust been established (usually via face-to-face interactions), then this 'relational proximity' becomes more important than spatial proximity in shaping access (see 5.5 Networks). Relational proximity refers to the quality of interpersonal relationships and level of trust that has been established between actors (Bathelt, 2006; Murphy, 2006). In this view, the barriers to access are not spatial but rather relational; access via the Internet is equal or superior to face-to-face contact when the actors have already established a close working relationship.

The notion of relational access highlights some of the many elements other than spatial distance that affect accessibility. First, as illustrated by inequalities in access to employment, many factors besides the map affect the ease with which an individual can reach destinations; these factors include, for example, household responsibilities, access to an automobile, familiarity with the local language. Second, economic geographers have long recognized that access depends on more than physical access (the ease of reaching destinations) alone (Women and Geography Study Group, 1984). Some people cannot gain entry to certain places, such as some clubs, even if they can physically get there because access is denied on the basis of gender, race or income, for example. For people without health insurance in the US, access to medical facilities is limited by ability to pay. Third, although the Internet provides a form of access to many people, access to IT facilities and the ability to use IT effectively are extremely uneven across different groups of people and places, an inequality known as the digital divide. Warf (2001) describes the dimensions of this divide at global and regional scales, while Gilbert et al. (2008) document the nature and extent of this digital divide at finer scales. Finally, as described in the chapters 1.3 State and 4.4 Global Value Chains, institutional barriers to trade and other forms of economic exchange hamper or prevent access, whether physical or virtual. How best to incorporate these non-spatial factors and the use of IT into measures of accessibility remains a question of considerable significance for economic geography.

68

KEY POINTS

- Accessibility refers to the ease of reaching destinations. Over the past several decades, economic geographers have shifted from thinking about access primarily in terms of the distances separating people and places to thinking about access in terms of the ability to connect with information, knowledge and people.
- Access is a key driver of economic processes such as the growth and prosperity of regions and individuals' ability to reach paid work.
- Before IT, accessibility required some form of mobility. An important topic of investigation in economic geography currently is the changing relationship between accessibility-with-mobility and accessibility-without-mobility, that is access via IT. Also of interest to economic geographers is the relationship between relational access (access based on trust) and spatial proximity. Cultural and institutional factors, such as language, race and legal restrictions, also affect access.

FURTHER READING

For an example of the relationship between grounded and virtual accessibility in the context of the logistics industry, see Aoyama et al. (2006) and in the context of individual daily travel, see Schwanen and Kwan (2008). Malecki and Wei (2009) discuss the emerging geography of submarine cables and their accessibility implications in a wired world. Transportation Research Board (2009) critically reviews the literature on accessibility and land use in the context of energy consumption in the US.

NOTE

1 Malecki and Moriset (2008) note, however, that spatial proximity still retains its significance in some industries, even mature ones such as garment manufacturing.

69

Section 3
Industries and Regions in Economic Change

Why do economies differ, and what makes an economy distinctive? In Section 3 we turn to four key concepts that are essential for understanding the role of industries and regions in economic change. Explaining the sources of economic variation across space has long been a disciplinary preoccupation of economic geographers and, as a result, various theories have been postulated to uncover how the invisible logic of the market manifests itself in economic landscapes.

The chapter on Industrial Location covers the work of many who are considered the founding fathers of the discipline, beginning in the nineteenth century. Often based on strict, and therefore unrealistic, assumptions, many early location models are increasingly disregarded in introductory economic geography classes. We believe, however, that every student who claims to know something about economic geography must know its disciplinary roots, which begins with the industrial location theories covered in this section. Furthermore, many of the basic logics and principles of classical industrial location theory still hold true and, as a result, they remain relevant to contemporary economic geography. Indeed, economic geographers who consider themselves to be regional scientists keep this tradition alive and are developing increasingly sophisticated models with advanced computing technologies, whereas other economic geographers have turned to increasingly cultural, social and institutional factors that are more difficult to quantify (see Section 5).

Understanding Industrial Clusters is popular among contemporary scholars and policy makers, but in this chapter we also show that the concept has deep historical roots. In this chapter we cover must-know economic concepts that are indispensable to understanding industrial agglomerations, the economic costs as well as the benefits accrued by such agglomerations and how these ideas have been translated into policy. We also examine why agglomerations have not disappeared with the centrifugal forces of globalization, and how economic geographers have generated theoretical frameworks to explain the simultaneous process of agglomerations and globalization.

72

Regional Disparity is among the oldest of policy concerns, and it remains a central motivation for economic geography as a discipline. Why is the distribution of wealth across space unequal, and why does this inequality persist? Regions are an important unit of analysis in understanding how cities and their economies are sustained and supported by their hinterlands. Unbalanced regional growth within a country has been viewed as indicative of economic inefficiency (i.e. resources are under-utilized) and social injustice. Major ideological debates have raged over the meaning of regional disparity; some analysts have seen disparity as a temporary condition that exists prior to equilibrium, whereas others have viewed disparity as a permanent and cumulative outcome of natural market forces. This debate has had important repercussions as many regional development policies have been implemented around the world based on these theories.

In contrast to the first three chapters in this section, Post-Fordism is a relatively new concept that goes back only to the 1970s. Interests in industrial organization, i.e. how firms and industries are organized to raise productivity, began in the 1960s when productivity of the US industrial sector was falling. American scientific management, which had dominated much of the twentieth century until then, seemed to have lost its magic touch. In examining the root causes of the 1970s economic crisis, economic geographers became intensely absorbed in the intricacies of epochal industrial regimes, as represented by Taylorism, Fordism and Post-Fordism. In addition, labour saw technological change and the rise of numerically controlled machines as threatening its current position. In the chapter on Post-Fordism, we examine the historical trajectories of dominant forms of industrial organization, based on empirical evidence primarily from the automobile industry.

3.1 INDUSTRIAL LOCATION

Early location theories of largely German origin from the nineteenth century are among the pillars of economic geography. In order to identify the economic principles underlying spatial structures, industrial location theorists sought to develop a generalisable framework that would explain why certain activities are located where they are. These theories assume rational actors ('economic men'), who want to maximise economic gain (profits) (see 1.2 Firm) and whose decisions are not affected by social and cultural factors. These assumptions allowed theorists to isolate the effects of transport and labour costs, even though outcomes of the models are somewhat unrealistic. Today, the basic premises of industrial location theories remain broadly applicable and relevant to our understanding the construction of the space-economy.[1] In this section, we examine major aims and assumptions of early location theories, as well as their later extensions.

Locational rent and land-use distribution

Early industrial location theory drew upon the principles articulated by Von Thünen (1826), who developed a model of agricultural land use showing that land-use variations can be explained by the distance from the market. The model assumes that: 1) there is an isolated state with a single central city (market) surrounded by agricultural land on a uniform plain, 2) farmers are rational profit maximisers who all face the same production costs and market prices, and 3) transport cost is proportional to distance. Through this model, Von Thünen developed the notion of economic rent (sometimes referred to as locational rent), which refers to the highest rent that the farmer can bid for a parcel of land (production and transport cost subtracted from the total revenue from crop sale at the market). The bid-rent curve illustrates how distance from the market determines the type and intensity of crops grown on a certain parcel of land (see Figure 3.1.1). If a crop yields high revenue

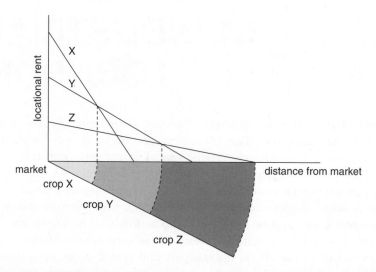

Figure 3.1.1 Bid-rent curve

Locational rents of three crops predicted by von Thünen model. Crop X is a perishable crop (perhaps tomatoes) and has a steep bid-rent curve, whereas crop Z is a less perishable crop (perhaps potatoes) and has a flatter bid-rent curve.

at the market but is perishable (and therefore entails a higher cost of transport), the bid-rent curve will be steep. Alternatively, if a crop yields low revenue at the market but is easily transportable (low cost of transport), the bid-rent curve will be relatively flat. When bid-rent curves of multiple crops are shown on the same graph, the highest-bidding crop will dominate the agricultural land use at a certain distance from the market. The model predicts a concentric agricultural land-use pattern with the market at its centre and with locational rents serving as payments for access to the centre. The bid-rent curve also shows the 'margin of cultivation' – after a certain distance from the market, the total transport cost becomes higher than the total revenue at the market and farmers have no incentive to produce any crop for the market.

The bid-rent curve has been used to describe the willingness of different urban land-use types to pay for accessibility, which is highest at the city centre (Alonso, 1964). In the urban case, the slope of the bid-rent curve for commercial land use is steepest, reflecting the fact that commercial establishments value accessibility (to potential customers) more than do industrial or residential land uses, and therefore are the highest bidders in the centre-city and dominate the landscape there. The industrial bid-rent curve comes next in its intensiveness, followed by the residential bid-rent curve. As a result, residential land use dominates the periphery of the city in Alonso's model.

Classical industrial location theory

Alfred Weber, brother of the famous German sociologist, Max Weber, is known in economic geography as the father of generalised industrial location theory. Alfred Weber (1929 [1909]) began the tradition of what is known as 'least-cost location theory' as his model aims at identifying the cheapest location for production. His work remains important as many principles of his location theory largely still hold today and explain broadly why firms locate their operations where they do.

Weber's industry location model presupposes a manufacturing plant to be located somewhere between the location of raw materials (R) and the market (M); the model takes into account wages and the cost of transport of raw materials to the plant, and the finished product to the market. The model assumes that the raw materials (R) needed for industrial production are only available in certain given locations, and the centres of consumption (M) are also given. Labour is also only available in certain locations, wages are given, and labour is considered immobile and unlimited in those locations. Furthermore, productivity is assumed to be uniform across the plain, as well as any other factors that may affect productivity, such as climate, culture, political systems and economic regulations. The costs of capital, rent, and equipment are also assumed to be uniform, and transport cost is proportional to distance.

The least-cost-location is then derived by calculating the transport cost of raw materials versus finished products (determined by weight and perishability) as well as low labour cost locations. According to this model, industrial plants in extractive industries (such as logging and mining) will be located near raw materials since the biggest

77

weight loss takes place during initial processing. In contrast, industries that produce highly perishable products (such as bread) should locate near the market, as transport cost is high for such goods. Within the 'critical isodapane' (the area in which total transport cost adds up to be the same), the plant should locate in places where wages are lower (Figure 3.1.2).

Weber acknowledged the presence of agglomeration economies and diseconomies, and incorporated these factors into his analysis (also see 3.2 Industrial Clusters). Agglomeration economies (cost savings from locating in close proximity to other plants) may come from sharing common resources such as a specialized labour pool or auxiliary services (accounting, financial services). Agglomeration diseconomies may come from traffic congestion or higher rent. Thus, plants are also more profitable when they agglomerate, as long as they are all within each of their critical isodapanes.

Among Weber's assumptions, the most problematic were that demand was held constant and labour was considered immobile. Stable demand meant that everything produced was consumed and that fluctuations of demand or decline in certain product categories were ignored. Also problematic is the assumption that labour location is not responsive to wage differentials; in reality, labour location varies across time and space. In the case of the US labour market, people are historically far more willing and likely to move long distances for jobs than in Western Europe. In addition, transport cost has declined significantly over the past century, making these models less applicable. Complicating matters further, many firms today are no longer single-establishment but maintain complex structures, in which manufacturing is conducted in low-wage countries and research and development (R&D) is located near important markets or corporate headquarters. Yet, for each establishment, Weber's principle may still broadly apply, and his influence in economic geography has been considered 'universal' (Holland, 1976).

Von Thünen's and Weber's ideas were extended by August Lösch (1954 [1940]), who synthesized their basic principles and drew on the work of Christaller as well (see also 6.3 Consumption). Employing the *ceteris paribus* (i.e. all other things being equal) assumption, Lösch defined optimal location as points where the difference between total revenue and total cost is the largest, while at the same time each producer maximizes its 'market area', which is laid out in a series of hexagons

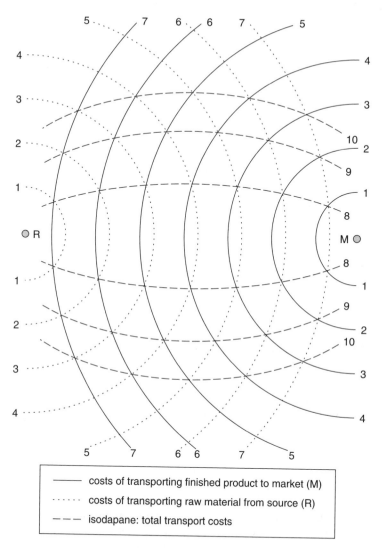

costs of transporting finished product to market (M)

costs of transporting raw material from source (R)

isodapane: total transport costs

Figure 3.1.2 Critical isodapanes
Assuming transport costs are proportional to distance and no weight loss occurs in transport, optimum production location falls between R (site of raw materials) and M (site of the market) where the total unit cost of transport is no more than 8.

79

much like Christaller's central place theory (1933). For Lösch, cities are not markets where dynamic agglomeration economies are at work, but super-impositions of multiple market areas. Lösch thus developed a more comprehensive model for industrial location, one that acknowledged demand (prices) and cost fluctuations and which was able to account for retail sector location strategies.

While Lösch recognized the presence of various important factors that shape regional growth, such as regulations, migration and entre-preneurship, these elements remained largely outside his model. Lösch eliminated the locations of raw materials and markets, which were central to Weber's model, and assumed that both are distributed uni-formly across the terrain. The outcome is a model that is much lauded as the first in presenting a full general equilibrium solution but, at the same time, is heavily criticised for being divorced from reality. Lösch was highly deductive in his methodology as he sought to incorporate a general equilibrium framework and used a far more restrictive set of assumptions than those employed by Weber.

80 Regional science and the quantitative revolution

Walter Isard, known as the father of regional science, is a founder and major source of inspiration for the Regional Science Association. Isard (1956) sought to expand the theory of the space-economy, by incorpo-rating dynamic processes and by taking into account spatial elasticity. His initial motivation came from a critique of general equilibrium theory in international trade, which was, in his words, largely taking place in 'a wonderland of no dimension' (1949: 477). Isard (1956: 23) sought to elevate descriptive analyses of the evolution of a region to a generalizable theoretical analysis in order to 'bring the separate loca-tion theories into one general doctrine'. Isard's solution to Lösch's model, for example, was to develop an eclectic approach that in part depends upon a 'combined Thünen-Löschian-Weberian framework' (1956: 19). While few would dispute the importance of Isard's work, he is also widely criticized for using the highly unrealistic Löschian model as a point of departure and for imposing restrictive assumptions, such as the absence of scale economies. Isard's reliance on linear program-ming and input–output analysis[2] led some to criticize his work as being driven and limited by techniques, making Isard's model static at best and inoperative at worst. While Isard may have succeeded in aligning regional science more closely with orthodox economics, and generating a dialogue between the two disciplines, his work left many economic geographers dissatisfied, because his models left too much unexplained.

Location theory in its most quantitative form thrived in the 1950s and 1960s during the 'quantitative revolution' that engulfed human geography on both sides of the Atlantic and was led by William Garrison (University of Washington) and Peter Haggett (Cambridge, Bristol). Industrial location theory caught the attention of the emerging field of operations research, and various scholars experimented with single-facility and multiple-facility models of industrial location (see, for example, Kuhn and Kuenne, 1962; Hoover, 1967). Methodologies ranged from linear programming to stochastic models, with the former focusing on simultaneous resource maximizing and cost minimizing, and the latter adopting a behavioural approach, in which the assumptions based on the economic man was relaxed to allow for variations in locational decisions (see 1.2 Firm).[3] Stochastic models incorporate random probabilities in locational decision-making and are at the root of simulation models, including Monte Carlo simulation.

81

Ultimately, the limitation of techniques became the limitation of theoretical advance in industrial-location theories. Dissatisfaction with quantitative methodology arose as greater social awareness and political activism in the 1960s drove the shift from models based on 'economic man' to other alternatives. The discipline of regional science went into an intellectual cul-de-sac (Holland, 1976; Barnes, 2004). Today, aside from the widely held recognition that principles of classical location theory still have broad applicability, consensus also exists that the current methodological tools do not allow us to extend research too far beyond what we already know about industrial location. Instead, location theory has diverged into three distinct trajectories.

Beyond location theory

There are at least three extensions of location theory in contemporary research in economic geography. First, under the neo-classical paradigm, location theory has been revived as geographical economics, much by the efforts of international trade theorists such as Paul Krugman, who won a 2008 Nobel Prize for his work in reintroducing space to mainstream economics. By combining trade and location theory, Krugman's main contribution was on the spatial consequences of increasing returns, instances where more than a proportional increase in output is observed from a proportional increase in inputs. Krugman viewed increasing returns as an underlining cause of specialization, which in turn affects patterns of trade and locations of industries (see 3.3 Regional Disparity).

Second, major economic restructuring in the 1970s, most deeply experienced in the US and UK, prompted interest in a structural approach to the study of location. Restructuring was largely seen as a crisis of capitalism, and along with structural development theories developed by Frank (1967) and Wallerstein (1979) (see 4.1 Core–Periphery), the structural paradigm was increasingly adopted to understand broad regional growth and decline (see Massey, 1979; Storper and Walker, 1989) (see 3.3 Regional Disparity). However, the structural paradigm was quite distinct from industrial location theories of the past in that it focuses on causal power relations, not on the specific locational manifestations of economic activities.

82

Third, economic geographers, along with researchers from business and management, have conducted various studies that seek to better understand the patterns of foreign direct investment (FDI) and locational choices of multinational enterprises (MNEs), using theories of international business, institutional economics and organizational behaviour, including the International Division of Labour, product life cycles (see 4.2 Globalization). These studies speak to our contemporary interest in the geography of off-shoring and uneven job creations subnationally as well as internationally.

In addition, many new factors have been identified that drive industrial location today. Labour is viewed to be vitally important, not just in terms of wages and quantity, but in terms of skills and quality (see 6.1 Knowledge Economy). Some argue that today, businesses increasingly follow workers, and particularly so in those industries seeking high-skilled workers of the industrialized world. These workers are attracted to locations that offer *amenity*, such as climate, culture, leisure and entertainment opportunities in addition to employment opportunities. Technological advancement, which has made possible the long-distance coordination of activities, has also made innovation activities highly complex. This gives further advantages to agglomerations, and particularly those that are in close proximity to government research laboratories and universities (see 3.2 Industrial Clusters).

KEY POINTS

- Classical industrial-location theories involve many unrealistic assumptions, yet many of the general principles laid out by the theories still stand today. Such phenomena as off-shoring and industrial clusters can be broadly explained by classical theories.
- Regional science as an extension of industrial-location theories, however, faced much difficulty, and some argue that the field of regional science has lost intellectual currency within economic geography. Methodological limitations and increasingly proprietary business and market data contribute to this difficulty.
- Today, interests among economic geographers have shifted increasingly to unquantifiable factors that explain industrial location, including regional culture, personal and business networks, and the role of skilled workers in innovation. Some argue that people no longer follow jobs, and instead jobs follow people, while others contest this view.

83

FURTHER READING

For a historical overview and contemporary extension of location theory, see Brülhart (1998). For a contemporary interpretation of location theory from the perspective of trade theory, read Krugman (1991). For an economic geographer's assessment of global industrial location, see Taylor and Thrift (1982) and Dicken (2007). See Storper and Scott (2009) for a view that industrial location is still largely determined by people following jobs.

NOTES

1 The English translation of the German word Raumwirtschaft and refers to the spatial structure of an economy associated with the German school of location theory (Barnes, 2000).

2 Input–output analysis is useful in specifying inter-sectoral supply and demand effects but is powerless in regard to incorporating scale economies and innovation, the two critical factors that are associated with industrialization and subsequent economic growth.

3 Allan Pred (1967), for example, developed a behavioural matrix for industrial-location decisions, in which he adopted the satisficer in place of maximizer principle.

84

3.2 INDUSTRIAL CLUSTERS

Industrial agglomerations (also known as industrial districts or clusters) are geographic concentrations of economic activities. These agglomerations have been traditionally understood as outcomes of the economic savings made possible through reductions in the average costs of production or service provision due to spatial proximity. These savings are known as external economies of scale. Various typologies of industrial districts exist in the literature, although these typologies should be regarded as ideal types. Industrial agglomerations are simultaneously an aspect of uneven development and specialization, and serve as important locales of job creation and innovation.

Agglomerations: economies and externalities

Fundamental to an understanding of industrial agglomerations is the notion of external economies of scale. Economies of scale refer to cost savings in per-unit inputs when certain scale is achieved, and can be internal or external. When economies of scale are external, cost savings are experienced collectively by firms in the locality, typically through the sharing of benefits from infrastructural development and the emergence of a specialized labour pool.

External economies of scale are considered in part generated by positive externalities. As one of the three conditions of market failures (others being monopoly and public goods), externalities are costs (negative) or benefits (positive) accrued to an entity above and beyond its accounting. An example of negative externalities, which would lead to agglomeration diseconomies is the air pollution produced by a factory (assuming no government regulations). An example of positive externalities, which would lead to agglomeration economies, is the sharing of factor inputs (e.g. a pool of labour with particular skills which would drive down search costs). When positive externalities outweigh negative externalities, industrial agglomerations will grow, taking advantage of agglomeration economies until diseconomies set in to prevent further growth. Industrial agglomerations can be reinforced by expanding the 'roundaboutness' of production[1] (Young, 1928; Scott, 1988).

There are two types of agglomeration – one that benefits from urbanization economies and another that benefits from localization economies. Localization economies result in agglomerations that specialize in one industrial sector (Marshall, 1920 [1890]). Urbanization economies refer to the advantage of being located in a large urban area, with large and heterogeneous markets (Hoover, 1948). This advantage is also referred to as Jacobian externalities, primarily born through the diversity of activities and the size of market combined generating opportunities to encounter the unknown (Jacobs, 1969). Florida (2002) drew from Simmel (1950) and argued that diversity in cities fosters creativity.

In contrast, localization economies refer to benefits that almost exclusively impact a particular industrial sector, leading to industrial districts. It is agglomerations of this type that have preoccupied economic geographers, as they are seen as places that induce technological innovation and productivity growth, ultimately making them internationally competitive and economically resilient. Because of the growing recognition of the importance of innovation in spatial organisation of firms, the role of technological spill-over as a critical externality in contemporary agglomerations is the subject of much research today (see 2.1 Innovation).

Marshallian agglomerations and industrial districts

Marshall (1920 [1890]) focused on the benefits to individual plants typically at the intra-sectoral level, to share common pools of factor inputs, such as land, labour, capital, energy, sewage, transportation, ancillary services and production know-how. The critical mass of such common pools, followed by specialization of these inputs will, in the long run, raise plant productivity, creating a district-specific 'industrial atmosphere' that is 'in the air', which is neither easily moved nor replicated. Marshallian agglomerations such as the textile industry in nineteenth-century Lancashire, England, are characterized by horizontal and vertical coordination among small firms that specialize in one function in an industry (e.g. dye-making or thread manufacturing). Small firms compete with one another in a purely neo-classical sense, respond primarily to price signals and they remain at arms-length relationships. Those who supply intermediate goods are typically oriented toward the local market, although final products may be exported. Aside from the market which serves as a final destination of products, it is implicitly assumed that Marshallian industrial districts have minimal interactions and linkages with those outside the district.

The renewed interest in industrial agglomerations was generated by the flexible specialization debate, in which Piore and Sabel (1984) advocated an industrial organization comprising territorially embedded small and medium sized enterprises specializing in one aspect of production, cooperating with each other, and engaging in small-batch, demand-pull production[2] (see 3.4 Post-Fordism). This offered a glimmer of hope for industries and regions that were increasingly faced with deindustrialization, offshoring and plant-closures.

Scott (1988) argued for a complementary analytical apparatus to the locational processes and outcomes of changing regime of accumulation (see 3.4 Post-Fordism) by bringing in transaction costs analysis from Institutional Economics (see 1.2 Firm). New industrial spaces are characterized by intense agglomeration economies and occur under the new regime of flexible accumulation. Scott points out that whether firms will choose vertical integration or disintegration depends on the combined effect of scale and scope on their production costs. Vertically integrated production would lead to company-towns, whereas vertically disintegrated production would lead to industrial districts.

87

Italianate industrial districts

Whereas Marshallian agglomerations rely on a neo-classical economic conceptualization of firms as collections of atomistic competitors, the Third Italy industrial districts emphasize interdependence among firms, cooperative competition and the importance of trust among economic actors within the districts (Bagnasco, 1977; Brusco, 1982; Russo, 1985). The Third Italy model of industrial districts is distinct from the Marshallian industrial districts primarily in the degree of coordination and cooperation among firms, or 'collectivization of governance' according to Amin and Thrift (1992). Firms in the Third Italy model achieve a much higher degree of coordination among small firms, typically based on craft form of work organization for garment and shoe-making industries.

Although competition exists, the firms tend to be highly specialized and this mitigates direct competition. Non-corporate entities, such as industrial co-operatives, coordinate various activities that benefit firms in the district, including developing shared auxiliary services, coordinating infrastructural development, sponsoring trade fairs, offering worker training programmes and providing capital in forms of loan guarantees.

Harrison (1992) contributed further insights on the debates over Marshallian vs. Italianate industrial districts, by claiming that the debates in the 1990s offered a new interpretation which acknowledged the significant role of social factors in geo-economic processes. Taking Granovetter's notion of embeddedness, Harrison emphasized the importance of trust, which is nurtured over time through formal and informal personal contacts, and geographical proximity is critical to build trust in the new interpretations (see 5.4 Embeddedness). Personal contacts developed through various social relations (including kinship and friendship), and nurtured through common religious beliefs or alliances in local politics expressed at meetings in social clubs and churches, complement business relations within industrial districts.

88

Geography of innovation

Innovative milieux represents simultaneously a type of industrial district and a research framework with a focus on locations of innovation. The GREMI (Groupe de Recherche Européen sur les Milieux Innovateurs) group in Paris was among the first to present the thesis that certain local industrial districts qualify as innovative milieux. As part of the GREMI group, Aydalot (1986) was among the first in advocating for an analysis of districts that are self-reproducing in the manner of 'creative destruction'. The GREMI considered innovative milieux as space shaping space ('active space' approach) and focused on the role of endogenous technological trajectories in shaping spaces of production (Ratti, 1992).

Economists have focused on the role of technological (or knowledge) spillovers in the clustering of innovation (Glaeser et al., 1992; Feldman, 1994; Audretsch and Feldman, 1996). Technological/knowledge spillover is a specific type of externalities (also known as Marshall–Arrow–Romer (MAR) externalities), and refers to knowledge generated within an agglomeration above and beyond what is contained within a firm, which is assumed to be captured best by the local producers. It should be noted, however, studies exist that dispute the strength of MAR specialization externalities over Jacobian diversification externalities (see Feldman and Audretsch, 1999; Harrison et al., 1996; van Oort, 2002).

Both Marshallian agglomerations and Italianate industrial districts are conceived to have grown organically, without state support. They are contrasted to districts that are primarily developed and strongly driven by state mandates, such as technopoles and science cities. Examples of the latter include Hsinchu Science City (Taiwan), Sophia Antipolis (France) and Tsukuba Science City (Japan) (see Castells and Hall, 1994). Furthermore, industrial districts are not confined to the manufacturing sector. For example, Hollywood has been analysed using the approach of industrial districts, and so has the City of London and its financial district (see Clark, 2002; Scott, 2005).

In comparing two industrial districts in the electronics industry, Silicon Valley and Boston's Route 128 region, Saxenian (1994) found that distinctive regional culture (see 5.1 Culture) plays a key role in promoting active entrepreneurship (see 2.2 Entrepreneurship). These regional institutions are embedded in specific contexts and nurtured over time, provide not only support and incentives, but also moral support to entrepreneurs. Saxenian argues that Silicon Valley is unique in accepting entrepreneurial failures as positive experiences, and the region attracts a number of 'serial' entrepreneurs who launch new businesses multiple times in their lifespan.

89

The most recent incarnation of industrial agglomeration is regional innovation systems (Cooke and Morgan, 1994; Asheim and Coenen, 2005). Asheim and Coenen (2005) defined regional innovation systems as 'institutional infrastructure supporting innovation within the production structure of a region' (p. 1177). Regional innovation systems can be further categorized into territorially embedded, regionally networked, and/or regionalized national innovation systems, by different types of knowledge used in their innovation systems, either analytical (science) or synthetic (engineering). These territorially based views of innovation encourage policy makers to incorporate geographic components in innovation policy.

Porter's industrial clusters

The term industrial clusters was popularized by an American scholar and a management guru, Michael Porter (2000). Porter defined clusters as 'geographic concentrations of interconnected companies, specialised suppliers, service providers, firms in related industries, and associated institutions (e.g., universities, standards agencies, trade associations) in a particular field that compete but also cooperate' (p. 15). He also calls them 'critical masses of unusual competitive success in particular business areas.' (p. 15). Clusters define microeconomic business environment, which in turn affects innovation, productivity growth and new business formation. Porter also points out that whereas some clusters are hotbeds of innovation and grow, others may suffer from a lock-in and face decline (see 5.4 Embeddedness).

Porter's sources of locational competitive advantage is based on an elaboration of his 'diamond' framework, which involves factor (input) conditions, demand conditions, related and supporting industries, and context for firm strategy and rivalry. These four factors interact to shape an industrial cluster. Economic geographers have viewed Porter's contribution as more pragmatic than theoretical, as his work is not characterized with new propositions. Instead, his contribution is seen as lying in his ability to convince policy makers as well as corporate managers of the importance of agglomerations. Porter argued that the cluster as a unit of analysis allows us to devise better policy for international competition, claiming that, 'the health of the cluster is important to the health of the company' (p. 16). Questions remain, however, on the novelty and originality of Porter's cluster theory. For example, Martin and Sunley (2003) argued that Porter's theory on clusters is more an analytical creation than reality, and accuse Porter of eclecticism, lack of specificity (particularly social dimensions) and context (situating clusters in the broader dynamics of industry and innovation).

Local buzz, global pipelines

Storper (1995) distinguished between the characteristics of localised capabilities and regional specific assets as traded and untraded interdependencies, with the former emphasizing the material trade and the latter informational trade. Storper (1997) proposed a multi-dimensional analysis that incorporates entrepreneurship and flexibility (local definition of the common good), labour conventions, vertical and horizontal inter-firm relations and its nature of competition (openness vs. membership) as well as its nature of relations (reciprocity, social change). According to Bathelt et al. (2004), buzz refers to spontaneous and fluid

'information and communication ecology created by face-to-face contacts, co-presence and co-location of people and firms within the same
industry and place or region' (p. 38). Through co-location, firms are able
to understand the local buzz in a meaningful and useful way by sharing
the common language, technology and attitudes, and build institutions
that reflect the particular 'communities of practice' (Brown and Duguid,
1991; Wenger, 1999; Amin and Cohendet, 2004) (see 5.4 Embeddedness).

Storper and Venables (2004) emphasized the importance of 'local
buzz' created by face-to-face contacts, which they argued involves
unique behavioural and communicational properties and therefore
offers specific advantages over other modes of communication. Moreover,
face-to-face contacts are an essential mode of learning-by-doing, or the
transfer of tacit knowledge (see 2.1 Innovation). Because of these
unique properties of the face-to-face contacts, agglomerations remain
resilient features of spatial organizations of economic activities even
after the broad proliferation of the internet (Leamer and Storper, 2001).

Markusen (1996) and Amin and Thrift (1992) criticized scholars' preoccupation with the local in the debates on industrial districts and
instead emphasized the role of global networks in sustaining industrial
districts. Markusen identified various typologies of industrial districts,
and among them, proposed 'hub-and-spoke' industrial districts, which
combine local strengths with external linkages. Amin and Thrift called
them 'Marshallian nodes', which serve as a centre of social interactions
in a global system. These nodes survive and thrive because of their
multiple roles – not only out of a symbolic value of belonging but also
in their knowledge base and their ability to offer access to firms and
markets outside the districts. Amin and Thrift argued that localization
of production and vertical disintegration goes hand in hand with the
globalization of networks.

Bathelt et al. (2004) focused on the distinctive properties and complementarities between local and external sources of knowledge, and
argued against a simplistic understanding of equating local buzz with
tacit knowledge and global pipelines with codified knowledge. Instead,
they argue that extra-local flows of knowledge, 'global pipelines', can
play a pivotal role in injecting new knowledge to local routines, and
thereby raise the innovative capacities of firms. Thus, firms reap the benefits of both local and extra-local knowledge to survive and thrive in the
contemporary globalized economy, and they are highly complementary
(see 2.1 Innovation; 5.5 Networks).

91

KEY POINTS

- Industrial agglomerations, districts and clusters are key preoccupations of economic geographers. These signify areas of specialized economic activities that induce technological innovation, productivity growth and ultimately successfully withstand international competition.
- Various types of industrial agglomerations exist, with external links of varying significance. The reasons for agglomerations also range from sharing of common factor inputs, social relations, to regional institutions and culture.
- Whereas industrial agglomerations, districts and clusters are by no means obsolete, the Internet has fundamentally altered the possibility of technological innovation and transfer. This is because face-to-face contacts still remain a far superior mode of communications, particularly in learning-by-doing and the transfer of tacit knowledge.

FURTHER READING

For a contemporary overview of clusters, see Bathelt (2005). More recently, Storper (2009) proposed to analytically distinguish various forms of agglomerations. For a critique of Tom Friedman's work *The World is Flat*, see the special issue of *Cambridge Journal of Regions, Economy and Society* (2008) Vol. 1, No. 3.

NOTES

1 It refers to the increasing time and complexities in labour process required to bring a product to the market.
2 Demand-pull production is contrasted to supply-push production. In demand-pull production, products are made after the orders are placed.

3.3 REGIONAL DISPARITY

What are regions, and what explains the differences in their economic activities? In the mid-1950s, North (1955, 1956) and Tiebout (1956a, 1956b) declared that there is no 'ideal' region, a view that still holds true in the early twenty-first century. Jacobs (1969) conceptualized a region as a self-sustaining entity comprising cities and the hinterland, with the former serving as a location of markets and exchanges while the latter provided agricultural commodities necessary to sustain the city. Today's conceptualization of regions has departed from the natural resource-based paradigm of the past to those that are based on other indicators, defined by export base, transport accessibility, intra- and inter-firm relations, as well as terms of regional and international trade.

What is a region?

A dominant conceptualization of the region today corresponds to the metropolitan area, consisting of places linked together via daily commuting and business-to-business relationships. In Europe, regions correspond with political/administrative units that are primary entities of regional development. Because inter-regional disparity is among the foremost policy considerations in the European Union (EU), research on regions remains active in Europe, especially as the EU continues to add additional countries. In the US, however, regional policy has had a more challenging fate, as functional and administrative regions do not always correspond. This lack of correspondence poses challenges for coordinating development across city and county boundaries and often results in political competition and lack of resources. A few exceptions include Appalachian Regional Commission, which coordinates regional development initiatives at the supra-state level, and Metropolitan Planning Organizations (MPOs), which are federally mandated to coordinate transportation planning for every metropolitan area.

Regions can also refer to supra-national areas; for example, the Growth Triangle centred around Singapore includes the neighbouring state of Johor Bahru, Malaysia, and Batam and Bintan Islands, Indonesia. Multinational corporations (MNCs) are drawn to this region as they take advantage of wage differentials and locate headquarter functions (Singapore), mid-level manufacturing (Johor Bahru) and low-end manufacturing (Batam) within close proximity. Yet others refer to regional trading blocs (NAFTA, EU, etc.) as regions. These regional trading blocs have preferential agreements for intra-regional trade, while coordinating policies for inter-regional trade.

Regions have long been the key unit of analysis for economic geographers in understanding uneven development. As we shall discuss in subsequent sections, some believe that uneven development is not only detrimental to social justice, but also is an inefficient form of economic development, while others view uneven development as an inevitable process. Since the Industrial Revolution, we know that industrialization does not take place uniformly across the landscape. Regional inequality is measured in a variety of ways. One way is to measure factor endowments, such as investment and labour force. Another way is to measure economic outcomes, such as productivity and income distribution. To measure the distribution of incomes in a region or an economy, the GINI coefficient is the most commonly used indicator (see Box).

Gini coefficient

The Gini coefficient was developed by an Italian sociologist and statistician Corrado Gini. The index measures distribution of income across population, and falls between the value of 0 (completely equal distribution) and 1 (completely unequal distribution). Graphically, the coefficient is illustrated with cumulative proportion of an area's population on the X-axis and cumulative income on the Y-axis, and calculated by the ratio of the size of the area that falls between the 45-degree line showing a perfectly equal income distribution and the Lorenz curve, the actual income distribution (A), and the rest of the area that falls beneath the Lorenz curve (B). The larger the area, the higher the coefficient, indicating higher income inequality.

The United States has a high business start-up rate, low income taxes and a propensity to promote innovation. Yet, the distribution of income in the US is also highly unequal, with a higher GINI coefficient than that for many countries in Western Europe in the past several

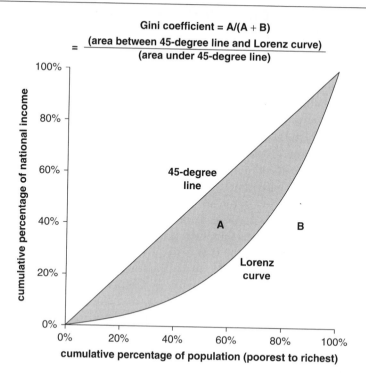

Figure 3.3.1 Gini coefficient

decades. Data also suggest that the US economy offers the highest executive compensation vis-à-vis average industrial workers' wages among various countries. In addition, inequality in the US, which was declining in the 1960s, made a U-turn in the late 1970s (Harrison and Bluestone, 1988).

Typically, the world's poorest countries are those that have persistently high GINI coefficients. However, in the past decade, GINI coefficients have risen dramatically in post-socialist countries (such as in Central and Eastern Europe) as well as in emerging economies (such as China and India). Some interpret these developments as increases in entrepreneurial opportunities and upward mobility that will eventually 'trickle down' to benefit the rest; others believe these trends will only exacerbate social divisions and lead to the destruction of communities. So far, no research has conclusively found one of these positions to offer a superior explanation.

Theories of inter-regional self-balance vs. self-imbalance

Theorists who have attempted to explain regional growth and disparity are broadly divided into two groups; those who believe that regions naturally balance out their growth prospects through perfect factor mobility between regions, and thereby minimizing inter-regional disparities (self-balance), and those who believe that regions naturally gravitate toward greater inter-regional inequality if they are left to market forces (self-imbalance). The former group assumes that firms maximize profits by locating in lowest-cost regions, and workers maximize their wages by moving to places with better employment opportunities (see 1.2 Firm, 3.1 Industrial Location). Ohlin's inter-regional trade theory (1933) claimed that imperfections in factor mobility, along with specialization (labour skills) that arises within different regions, would encourage interregional trade, and ultimately result in equalization of prices across regions in the long-run. In Ohlin's mind, factor mobility and trade are substitutes for each other, and the presence of spatial frictions is the cause of inter-regional trade.

While self-balance theorists view regions as parts of an idealized capitalist system without barriers to factor mobility, self-imbalance theorists understand regional disparity as being intrinsic to capitalist economic growth. Regional self-imbalance occurs because of asymmetry and limits in capital and labour mobility. Adopting a historical-structural view, these theorists observed that capitalists' pursuit of efficiency, combined with technological advances, had transformed previously dispersed, cottage-based and crafts-driven production into large factories with scale economies. In their view, a cycle of capitalistic accumulation, of profits, investments and job creation, induced largely involuntary rural-to-urban migration, and created what Marx called the 'industrial reserve army'. Gunnar Myrdal (1957) described this process as circular and cumulative causation, a concept that grew out of his study of racial inequality in the United States. Constructing a framework that explained not only social relations but more specifically the economic transformation of regions, Myrdal emphasized that market forces tend to increase, rather than decrease the uneven geographical distribution of wealth under capitalism. Cumulative causation can set forth an upward spiral, which involves virtuous cycles of economic development through the multiplier effect. The multiplier effect is a process

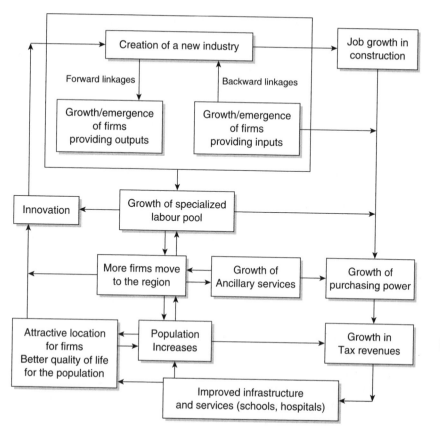

Figure 3.3.2 Circular and cumulative causation

whereby a new or expanded economic activity raises the purchasing power of households and adds more jobs, which in turn increases demand for housing and consumer goods and services, thereby creating opportunities for business expansion and further growth in employment. More people, more industries and improved infrastructure make the region attractive for businesses to move in. However, cumulative causation can also describe a downward spiral of economic development, leading to regional decline.

Uneven development occurs when one region attracts factor endowments (more capital or labour) at the expense of other regions that lose factor endowments that could have otherwise led to economic development.

This process was called 'polarization' by Hirschman and 'backwash' by Myrdal. However, counter-trends also exist, in which businesses seek out regions with lower wages, less congestion and lower rent. These counter-trends – known as 'spread' (Myrdal), 'spill over' (North) or 'trickle down' (Hirschman, 1958) – will eventually balance economic activities across regions. When the trend toward backwash or polarization dominates, regions become increasingly unequal in their economic activities. Alonso (1968) argued that the process of regional development takes place with successive waves of centralization and decentralization.

Krugman's (1991) seminal paper 'Increasing Returns and Economic Geography' reconnected many economists to the work of early location theorists (e.g. Christaller, Lösch, Hoover) but also combined other traditions in economics such as Myrdal (1957) and Arthur (1989) that reinvigorated geographical economics. Although historians and economic geographers have long known about manufacturing concentrations and what allowed them to emerge (i.e. mass production combined with improved transportation infrastructure), Krugman's main contribution was to develop a general equilibrium model based on the interaction of economies of scale with transportation costs. The model assumed two regions and two kinds of production, agriculture and manufacturing. The agricultural sector is subject to constant returns constrained by land requirements, transportation costs are assumed to be zero,[1] and the labour force is immobile. In contrast, the manufacturing sector assumes the presence of economies of scale, iceberg transport costs,[2] and a 'skilled' labour force that is mobile. Krugman argued that the economy of a region that combines high transportation costs and weak economies of scale will be based primarily in agricultural activities, whereas in a region that combines low transportation costs and strong economies of scale, circular causation will drive manufacturing concentration.

Furthermore, Krugman argued that workers concentrating in one of the two regions can be understood as an equilibrium, by incorporating the presence of pecuniary externalities, which arise out of demand or supply linkages within the manufacturing sector (i.e. one manufacturer's action affect the prices of goods produced by another manufacturer). Workers presumably take into account real wages (which are in part determined by pecuniary externalities in the manufacturing sector) as well as transport cost to decide if they will stay in their home region or relocate to the other region. Workers' decisions then set the boundaries between concentration and dispersion of the manufacturing sector. Krugman's main contribution was in his mathematical construction that explained regional disparity. Although the model was highly abstract and built upon heroic assumptions about firm and worker behaviour, his model was found to be compelling and has been extended by numerous scholars in geographical economics and regional science (e.g. see Lanaspa and Sanz, 2001; Combes et al., 2008).

Export-base theory of regional development

North (1956) developed export-base theory as a critique of economic-sector theory, developed independently by economists Clark (1940) and Hoover and Fischer (1949). According to them, the dominant sector of employment will shift from the primary sector (agriculture, forestry, fisheries, mining), to the secondary sector (construction, manufacturing, utilities), and finally, to the tertiary sector (business services, personal services, social services, retail and distribution, transportation and communications) along with the growth of real income per capita. The shift occurs as a result of the rise of labour productivity in agriculture (through mechanization), which offers labour surplus to other sectors of the economy. Labour productivity in industries will also rise, as a result of increased efficiency through the division of labour (specialization) and mechanization (and, ultimately, full automatization), releasing labour surplus to the service sector. The service sector absorbs much of the surplus labour, in part because certain services cannot be mechanized and labour productivity remains low.

99

Pointing out that the Hoover and Fischer thesis is inherently Euro-centric, North claimed that in North America, regional development has occurred primarily through exports, rather than as a gradual shift from a subsistence economy. This is particularly prominent in the Pacific Northwest, where the region grew by attracting outside capital and by exporting commodities, such as lumber, to distant markets. Thus North's export-base theory contends that regions can grow without industrializing, and by depending on external markets. North suggested that an export base can lead to diversification, and ultimately regional growth. North is particularly adamant that Hoover and Fischer's third stage is dependent on a whole variety of factors, from changing demand and new technological development (in transport and in production) to competition from other regions as well as government subsidies, and war.

Tiebout (1956a, 1956b) contested North's export-base theory by arguing that it applies only in a small region and in the short run. For Tiebout, it was not the export-base industry that induces regional growth but, rather, a strong 'residentiary' industry that reduced cost to make the export-base sector competitive, thereby allowing regional growth. However, North maintained that capital and labour infusion from outside the region (export base), government subsidy, or migration, can create the residentiary sector, through the process of import-substitution.

Regional development and policy

Economic-sector theory raises important questions about how policy makers set sectoral and geographical priorities. Priority sectors are typically those that are highly innovative, as they constitute the highest promise for expansion and productivity growth. Hirschman argued that priority sectors are those with 'thickest' input–output relationships, and particularly with significant backward linkages.[3] This makes sectors that have higher multiplier effects (e.g. automobile assembly) more valuable for the goal of sustainable regional economic growth. Kuklinski (1972) analysed international data that included North America, Western and Eastern Europe, as well as the Soviet Union, and claimed that chemical and metal industries are the priority sectors. In contrast, Kuznets (1955) emphasized the importance of innovation in regional development (see 2.1 Innovation). More recently, the effectiveness of service-sector led regional development has come to be questioned (see 6.1 Knowledge Economy).

Geographic priorities have also been considered in the regional development literature. Perroux (1950) developed the concept of 'growth poles' (*poles de croissance*) driven by a 'propulsive industry', a fast growing, dominant industry (or firm) with significant backward- and forward-linkages (Hirschman, 1958). Although the concept was initially non-spatial and referred to a set of industries or firms, it developed geographical connotation as it became an instrument of regional development policy. The policy was employed in France, where interests in balancing out concentration of economic activities and population in Paris metropolitan region were met with the policy of stimulating growth in other places, such as Toulouse, Strasbourg, Bordeaux, Marseilles and Lyon, which were chosen as locations for propulsive industries. Growth-pole strategies, however, tended to over-emphasize the benefits of external economies of scale at the regional level over internal economies of scale which operate at the firm level (see 1.2 Firm). Furthermore, government incentives alone have proven ineffective in attracting firms to otherwise undesirable locations, let alone generating self-sustaining growth.

Nonetheless, it is widely recognized that state actions have been critical in balancing the polarizing tendency of capitalist economic growth (see 1.3 State). The welfare states of Western Europe, for example, have invested considerable resources and planning regulations to help equalize economic growth across regions for much of the post-war period. The state can also facilitate growth in certain regions by concentrating resources. For example, although the United States traditionally shied away from intervening in private-sector activities through industrial-location policy, Markusen et al. (1991) make an intriguing suggestion that US defence policy (enacted through military installations, research laboratories and defence contracts to the private sector) contributed to the demise of the American Midwest and facilitated the growth of high-technology industries in California and New England. Castells and Hall (1994) document a wide variety of science-cities around the world, many of which are conceived and driven by the state (see 3.2 Industrial Clusters). Globalization of capital and the monopolistic power of multinational corporations (MNCs), however, may significantly reduce state power, or at least may lead to states competing for MNCs by offering them incentives (see 4.2 Globalization).

101

KEY POINTS

- The cause of inequalities between regions is an important starting point for theoretical discussions of growth and decline in economic geography. Such discussions also involve strong policy implications.
- Outcomes of the causes of inequality vary by the types of assumptions involved. Some underlining assumptions have included endogenous vs. exogenous markets, mobility of capital and labour, cumulative causations and inter-firm linkages.
- Debates on regional disparity have also focused on the need to develop a general theory vs. the need to explain reality. Economic

geographers have employed various concepts and tools, including neo-classical economics, historical (economic-sector) perspectives and Marxian social theory.

FURTHER READING

Holland (1976) provides a comprehensive overview of theories on regional self-balance and imbalance. For contemporary interpretations of regional development and policy, see Pike et al. (2006) and Hudson (2007). See the World Development Report (2009), which is based on the assumptions of regional self-balance.

NOTES

1 This assumption was made so that agricultural earnings are the same in the two regions.

2 Developed by Samuelson, a notable economist, 'iceberg transport cost' refers to the idea that a portion of the goods will 'melt' before reaching destination (therefore, transport cost is linear, based on weight, and is proportional to distance). The use of iceberg transport cost is controversial for economic geographers as it does not reflect reality.

3 As shown in Figure 3.3.2, backward linkages refer to linkages in production processes in the direction of raw materials, whereas forward linkages refer to linkages in the direction of the market. For example, for an automobile assembly plant, a linkage to a steel maker is a backward linkage, while a linkage to a car dealer is a forward linkage.

3.4 POST-FORDISM

The Fordism/Post-Fordism debate represents a periodized under-standing of the dominant mode of industrial organization and its role in economic growth, primarily in advanced industrialized economies. The term Fordism refers to streamlined assembly-line (mass-production) manufacturing with a strict division of labour, pioneered by Ford Motors of the United States. Fordism was the dominant model of high-efficiency production organization from its inception in 1913 to the end of 1960s, and is viewed as a major reason for the dramatic rise in pro-ductivity in the US manufacturing sector, making the United States economy an industrial powerhouse of the twentieth century (Best, 1990). Widely advocated by a group of French Marxian economists called the Regulation School and American labour economists in the 1970s and 1980s, Post-Fordism referred to the broad understanding that the end of the absolute superiority of the US economy and the advent of the subsequent crisis were in part explained by the end of the Fordist mode of production.

From handicraft to Fordist mode of production

Fordism emerged gradually, and was preceded by manufacturing through 'interchangeable parts' as well as by Taylorism; it was finally perfected by Henry Ford of Ford Motors in the form of the assembly-line of mass production. The manufacturing of interchangeable parts was made possible in improvement in machinery design for particular parts along with innovation in precision-measurement instruments, and was per-fected in armouries in New England by the mid-nineteenth century (Chandler, 1977; Best, 1990). This represented a revolutionary change, as the manufacturing sector had relied on skilled handicraft workers, who typically gained uncodified knowledge through years of apprenticeship about the entire process of production. Unlike such handicraft-based European practices, interchangeable parts achieved reproduction of a large volume of parts that are identical with absolute precision, so that any two parts can be assembled with-out hand-fitting by skilled handicraft workers. This process innovation had profound impacts on work organization; it increased the demand for skilled labour among for those who design and operate specialist machinery, and it created a demand for unskilled assembly workers (see 2.1 Innovation).[1]

Taylor (1911) was an industrial engineer who took part in the efficiency movement, and introduced a new management system which combined detailed time management for a worker's every motion, with careful coordination and planning in management, combined with an incentive wage system based on output volumes. In 1913, Ford Motors linked the American system of interchangeable parts with the streamlined flow of production in what we now know as the 'assembly line', and saw productivity increase dramatically.

The Fordist mode of production efficiently generated a massive number of identical products, and is therefore known as mass production. It thrived in an economic environment where demand was predictable and growing, and consumers were eager to fulfil their basic wants and needs, with relatively little interest in differentiation. In addition, the production organization under Fordism was particularly suitable for the American workforce and its diversity. For example, in 1915 Ford's Highland Park factory had 7,000 workers, most of whom came either from farms or from abroad; as a group they spoke more than 50 languages, and some knew very little English (Womack et al., 1990). Still they were able to produce complex products with high efficiency as work organization through task specialization required minimum training.

Furthermore, Ford opted to implement vertical integration by internalizing many parts of production for two reasons; first, his production was far more efficient than his parts manufacturers, and second, he was naturally distrustful in nature (Womack et al., 1990). As a result, for much of the twentieth century, American scientific management replaced European handicraft-based manufacturing as the dominant production system. The shift from cottage industries to large-scale, modern factories during the industrial revolution introduced technical and social divisions of labour, which had profound implications on regional economies.

Fordism and the economic crisis of the 1970s

By the mid-twentieth century, Fordism was the symbol of modern efficiency in American industrial organization. During the global economic crisis of the 1970s, however, when US industrial supremacy began to falter, it came under scrutiny for the first time. During this time, both UK and US economies suffered similar symptoms of deindustrialization, job-loss and stagflation.[2] Many indicators, such as productivity in the manufacturing sector, the US share of world trade, and the US share of multinational corporations (MNCs) showed downward trends (Bluestone and Harrison, 1982).

This led to a re-examination of Fordism as an effective production organization, particularly in view of changing macro economic environments, such as the end of the Bretton Woods Accord on fixed exchange rates and the gold convertibility of the US dollar (1944–1971), economic competition from Western Europe and Japan, and the emergence of Newly Industrializing Economies (NIEs),[3] as well as the two oil crises of the 1970s. The end of the Bretton Woods Accord in particular had significant disruptive effects (see 6.2 Financialization). For one, it prompted the emergence of regional trading blocs (such as the European Community – which evolved into today's European Union), and for another, discouraged investment in the Fordist mode of production, which depended upon a stable monetary environment (Piore and Sabel, 1984). This resulted in a worldwide recession in the early 1980s, with most western industrialized economies experiencing negative GNP growth.

105

The Regulation School

The French Regulation School was among the first to declare the end of Fordism. The School includes work of French macroeconomists starting from the 1970s. Aglietta's pioneering piece, entitled *A Theory of Capital Regulation* (1979[1976]), was followed by notable authors such as Boyer (1990), Coriat (1979) and Lipietz (1986). The Regulation School was an intellectual project of grand design, in which scholars attempted to develop a dynamic explanation of long-term structural economic transformation with a focus on social relations as endogenous factors that shape economic growth and crisis. Falling somewhere between pure theory and stylized facts (in their phrase, 'meso-theory'), the Regulation School exhibits a curious mixture of neo-classical economics and Marxism; it entails a unique framework that views the evolution of capitalism in terms of interactions between history and theory, social structures, institutions and economic regularities. Unlike neo-classical economics, in which institutions are typically viewed as barriers to the functioning of a perfectly competitive market, the Regulation School views institutions as building regularities that are conducive to growth, and can therefore have positive impacts on productivity. Unlike standard Marxism, the school viewed structural forces in the economy as naturally divergent, therefore requiring a non-deterministic analysis.

The Regulation School views class struggle as a primary expression of social relations, which also produces norms and laws, and the state is viewed as an expression of, and agent governing, social regulation (Aglietta, 1979 [1976]). The theory incorporates technical conditions of production and rules of distribution. The School claims that economic growth is achieved when there is cohesion between the regime of accumulation and modes of regulations. The regime of accumulation refers to systematic organizations of production, distribution and consumption through which productivity gains are extracted, shared and diffused. The mode of regulation refers to a specific local and historical collection of structural forms or institutional arrangements, which range from micro (e.g. labour contracts) to macro institutions (social security, international-exchange regimes, education). US dominance in the world economy was achieved through Fordism or mass production, a particular regime of accumulation and its associated mode of regulation that significantly boosted labour productivity and prompted a shift from an extensive to an intensive form of capital accumulation in the 1920s–1930s. However, once Fordism reached its technical and social limits in the late 1960s, it no longer yielded additional productivity gains, prompting the US economy to enter into structural crisis.

Although the Regulation School typically focuses on the national scale, Lipietz (1986) used the framework to promote understanding of the global political economy. According to him, modes of regulation, combined with external relations, explain the relative position of countries in the global political economy, and it was the spread of mass production beyond its original homelands ('peripheral Fordism') that created intensified international competition (Lipietz, 1986; Piore and Sabel, 1984). Therefore, unlike world-systems theory, the Regulation School incorporated systems of production with international power relations in order to better understand the world economy (see 4.1 Core–Periphery). In that sense, the Regulation School pre-dates, and is similar to, the global commodity chain approach (see 4.4 Global Value Chains).

The Regulation School has been criticized from a number of fronts. Some critics point to the ambiguity of concepts; others object to the somewhat rigid periodization as limiting the analysis. Although Aglietta (1979 [1976]) referred to neo-Fordism, Post-Fordism was never clearly articulated. Finally, the framework developed by the school is explanatory rather than prescriptive, and was limited in developing policy applications.

Flexible specialization and critiques

With backgrounds in political science and labour economics, Piore and Sabel (1984) contributed their seminal work *The Second Industrial Divide*, which was heavily influenced by the Regulation School as well as Italian scholars in regional development (Bagnasco, 1977; Brusco, 1982; Russo, 1985). While the Regulation School remained somewhat tentative in proposing the future of industrial organization (see discussion by Aglietta (1979 [1976]) on neo-Fordism), Piore and Sabel argued that one alternative to Fordist production organization is flexible specialization, which is characterized by small, specialized firms coordinating territorially based, small-batch and demand-responsive production. Taking the case of small Italian industrial districts of garment and shoemaking, Piore and Sabel showed that industrial survival is possible even in high-cost industrialized economies. In their view, flexible specialization reconciled competition and cooperation, promoted competition that induces innovation and developed community responses to price rigidities, a process that works toward long-term improvement in working conditions (see 3.2 Industrial Clusters).

107

The concept of flexible specialization led to active research on industrial districts, not only in Italian cities, but also in Silicon Valley (Saxenian, 1985, 1994; Kenney, 2000), Baden-Württemburg (Sabel et al., 1989; Cooke and Morgan, 1994; Herrigel, 1996) and Toyota City (Fujita and Hill, 1993), as well as non-industrial agglomerations such as Hollywood (Christopherson and Storper, 1986; Storper and Christopherson, 1987) (see 3.2 Industrial Clusters). The concept of flexible specialization also attracted a number of criticisms. Amin and Robins (1990), for example, believed that the structural change was far more complex and contradictory. The necessary institutions, the types of trust and social networks observed in the Italianate version of flexible specialization had not been observed in the United States. Also, some critics saw the idea as a nostalgic call for the return to neo-feudal cottage industries.

Also, by arguing for the benefits of flexible specialization, Piore and Sabel contributed to debates on the normative aspects of industrial organization, particularly those associated with labour management. For example, Bluestone and Harrison (1982) focused on 'locational' flexibility, observing that firms were becoming increasingly multi-establishment and could therefore undermine labour interests by simply threatening to relocate. Harrison (1994) went further by criticizing the euphoric optimism associated with flexible specialization and pointing to its potential 'dark sides' (see 5.4 Embeddedness). In particular, the flexiblization of labour has arguably reduced job security, through the increased use of part-time, contract (through temporary services agencies), or self-employed workers (Peck and Theodore, 1998) and as a result of declining labour unions (Rutherford and Gertler, 2002) (see 1.1 Labour).

The exchange between Gertler (1988, 1989) and Schoenberger (1988, 1989) exemplifies the liveliness of the flexible specialization debate at the time. These authors scrutinized the concept for its conceptual clarity, predictive ability and the ability to interpret geographic processes. Schoenberger emphasized the geographic dynamics that underlie the demise of Fordism, and proposed to include Western Europe as the Fordist core. In doing so, she demonstrated how Fordism relies on the maintenance of an orderly oligopoly, which in turn hampers the development of more promising competitive strategies at the core.

Gertler questioned the conceptual validity of Post-Fordism and, in particular, whether the alleged end of Fordism represents a distinct break with the past practices and geography of production. In Gertler's view, flexible specialization not only requires close, collaborative interfirm relations, but also may increase reliance on labour through the process of reskilling. He also reminded us that flexible institutional arrangements like subcontracting or the flexible use of labour pre-dated the alleged arrival of post-Fordism.

Lean production

Womack et al. (1990) added another empirical dimension to debates on emerging modes of production. Through a study of Toyota Motors in Japan, these authors show that the success of a firm can be largely attributed to lean production. Otherwise known as just-in-time (JIT) production, Toyota used small-batch production of different models, rather than emphasizing volume production of identical models. It also minimized business risk by drastically reducing otherwise costly and space-hogging inventory. It reorganized the labour process, replacing single-task with multi-task workers. Finally, it encouraged team-work among workers, which contributed to a sense of ownership in the products and ultimately to fewer defects.

Lean production allows a variety of models and options within a single assembly line (e.g. producing automobiles). As a result, it is suitable in an economic environment of uncertain and/or quickly changing demand, and for products that require customization for niche markets. Small-batch production minimizes business risk by adopting a demand-pull model, as opposed to supply-push model. It requires close coordination with suppliers, however, as parts must also be delivered just-in-time in small batches. It also requires significant worker involvement, and worker willingness to participate in longer training. The introduction of flexibility into an otherwise rigid Fordist mode of production was revolutionary in the 1960s, although this model has been widely adopted subsequently and has become a global standard for not only automobile production, but also for other products, such as personal computers.

109

Beyond Post-Fordism

The debates over new forms of industrial organization, represented by the Regulation School and flexible specialization, are important not only for their historical significance, but also for their impact on the direction of economic geography. As noted in articles published by *Economic Geography* at the time, the focus of the sub-discipline increasingly shifted away from traditional urban economic topics (e.g. problems of transport accessibility and the structure of metropolitan areas) to industrial and corporate geography. The geography of enterprises (the study of firms, mostly in manufacturing and particularly in high-tech) and industrial organization dominated the sub-discipline for much of the 1980s and the first half of the 1990s. Although the Fordist/Post-Fordist framework still retains strong currency, it has been criticized as too rigid and dualistic, too skewed in its focus on intra-firm or intra-cluster organizational dynamics, and limited by its almost exclusive focus on advanced industrialized economies.

If there is no consensus on the demise of Fordism as the most competitive industrial organization, few scholars agree on the current nature and state of post-Fordism. Today, our understanding of post-Fordism incorporates aspects of industrial organization discussed by the French Regulation School, as well as by proponents of Flexible Specialization and Lean Production.

KEY POINTS

- The French Regulation School developed a sophisticated conceptual framework that integrated micro-level practices on the shop-floor with macroeconomic trends.
- Piore and Sabel extended the discussion and suggested that Post-Fordism would be dominated by a production system known as 'flexible specialization'.
- No consensus currently exists, however, on the specificity of Post-Fordism regarding its production regime and its geographic consequences.

110

FURTHER READING

For a comprehensive overview of the Regulation School, see Boyer (1990). Also, both Dunford (1990) and Jessop (1990) offer a view of the Regulation School.

NOTES

1 Historians have disputed whether this system has led to the overall diskilling of labour. See Best (1990) for details.
2 Stagflation refers to the simultaneous increase of unemployment and inflation rates. Until the 1970s the Phillips Curve, which was developed based on evidence from the UK in 1861–1957, had us believe that unemployment and inflation were always inversely related, and many European economies justified their high unemployment as a way to control inflation.
3 The original NIEs (or sometimes called NICs, Newly Industrializing Countries) were: South Korea, Taiwan, Hong Kong and Singapore. They were also known as the 'Four Asian Tigers'.

Section 4
Global Economic Geographies

How do economic geographers conceptualize economic processes at the global scale? In Section 4 the focus shifts from the local and regional to key concepts that are essential to analyzing Global Economic Geographies.

How do economic geographers engage in globalization debates? As economic geographers are concerned with generalizable theory as well as specific contexts, the approach is necessarily multi-faceted. The chapter on Core–Periphery draws heavily on Marxist ideas, especially dependency theory and world-system theories, to explain how uneven economic development at the global scale is sustained in part by political-economic relationships between wealthy (i.e. core), emerging (i.e. semi-peripheral) and poor (i.e. peripheral) countries. Core economies, through colonization and imperialism and, more recently, neo-liberal and information-driven capitalism, have created uneven structural relationships with those in peripheral and semi-peripheral countries.

These relationships enable the core to extract surplus value and maintain global economic dominance, while making it difficult for those in the periphery and semi-periphery to improve their economic and geopolitical position in the world system.

The chapter on Globalization focuses on three aspects of economic globalization. The first is the role of MNCs in driving various forms (e.g. offshoring, outsourcing), which in turn result in a new International Division of Labour. We then discuss how globalization's interconnecting and homogenizing tendencies are limited by factors such as localized, place-specific knowledge, state actors, and cultural preferences and traditions. Finally, we explore the concept of scale under globalization, noting that traditional perspectives on scale (i.e. local, regional, national and global) are losing their usefulness for analyzing both local development and global economic change.

The chapter on Circuits of Capital revisits a Marxian interpretation of uneven development, but with a focus on how flows of resources (e.g. commodities, money) shape interactions among cities, regions and the global economy. The chapter begins with a description of Marx's three primary circuits of capital flows – money capital, productive capital and commodity capital – and explains how the circuits help to ensure that capitalists can continually extract surplus value from labour. This foundation is then extended through a discussion of David Harvey's seminal work *The Limits to Capital* which links capital circuits to urban and regional development processes. Harvey's work provides a framework for understanding how local development outcomes are significantly influenced by global flows of capital that serve as sources for inward investments and as outlets for the distribution of locally generated profits.

Finally, the Global Value Chains (GVC) chapter examines global economic geographies through the transnational supply chains linking commodity and component suppliers, manufacturers/processors, retailers and consumers. Starting with the global commodity chain (GCC) concept, first developed by economic sociologists and world-system theorists, the chapter traces the evolution of GVC research. Particular emphasis is placed on two key concepts in value chain relationships – governance and upgrading. Governance provides an essential means through which buyers and producers organize supply-chain relationships and control lower-tier suppliers with the goal of reducing costs and maximizing profits. Upgrading refers to the prospects for lower-tier suppliers to improve their position in the GVC through innovations that increase the value of their products, improve the efficiency of their operations and/or enable them to perform new functions (e.g. design, marketing). The chapter concludes with a discussion of the increasingly significant role of international standards (e.g. ISO 9000, product certification) in GVC governance and with insights from economic geographers on how localized phenomena (e.g. clusters, institutions) shape the prospects for, and significance of, value chain integration.

113

4.1 CORE–PERIPHERY

Why do some regions and countries remain chronically poor while others continually get richer and more powerful? Why has inequality increased in the past century and why is it so difficult for developing nations to consistently grow their economies and become more industrialized? The core–periphery concept addresses these questions by providing a framework for understanding how national and regional economies develop in part through their relationships to other places and regions. In doing so the concept helps to explain why there are 'have' places (cores) and 'have-not' places (peripheries) in the world economy. Economic geographers have extended the concept by applying ideas from radical political economy to explain how structural inequalities between core (advanced industrial) and peripheral (developing) countries have evolved and been sustained over time.

Dependency theory, world-system theory and the core–periphery concept

The idea that there are cores and peripheries initially came from the work of radical political economists (i.e. those inspired by classical Marxian theories on class relations and uneven development) interested in understanding how historical and political forces create unevenness in the world economy. Scholarship in this tradition originated in the 1950s and 1960s through critiques of mainstream theories about economic growth and modernization, particularly the work of Walter Rostow. Rostow (1960) argued that all countries that adhere to liberal-democratic principles and foster a free-market system can develop independently through five stages of economic growth.[1] Radical scholars disputed Rostow's model because it ignored the impact of colonialism, imperialism and contemporary geopolitical relationships on developing countries. From these concerns and critiques came dependency theory and world-system theory.

Dependency theorists' critiques of mainstream development theory centred on explaining how surplus value transfer and unequal exchange structure the relationships between countries once colonized (i.e. the periphery) and those responsible for colonization (i.e. the core). Baran

(1957) argued that the colonial experience in Latin America and other peripheral regions (e.g. Africa, South Asia) had created extreme levels of inequality that structurally separated the rich (bourgeoisie) from the poor (peasants) within those former colonies. Because of this inequality, the rich would have no incentive or interest to improve social and working conditions for the peasantry. This lack of social investment kept wages low, made commodity goods cheap in core export markets, and ensured that surplus profits were continually transferred to the local rich and/or the foreign investors who owned factories, plantations, forests and mines throughout the periphery.[2] Given these realities, the capitalist system linking metropolitan states (core) and satellite economies (periphery) would perpetually disadvantage the periphery vis-à-vis the core (Frank, 1966).

Emmanuel (1972) extended these ideas and argued that the unequal exchange between core and periphery is driven principally by wage disparities. Factory workers in the core, better organized and in relatively short supply, could demand and receive higher wages than those in peripheral economies where a surplus of unskilled labour and repressive political systems kept the working classes disempowered. Conditions for workers in the periphery were made even worse by colonial and post-colonial policies that promoted plantation-style agriculture, large-scale natural resource extraction industries, and the creation of development 'islands' where the workers were mostly male and employed in formal sector activities (Taylor and Flint, 2000). Because their families lived separately, the costs of social reproduction (i.e. child-bearing and child-rearing activities) were externalized to isolated rural areas where women and children lived on wage remittances from husbands and subsistence agriculture. As Figure 4.1.1 demonstrates, the low wages given to male workers in the periphery, coupled with the externalization of the social-reproduction process, created 'semi-proletarian' households that allowed workers and capitalists in core economies to benefit from surplus transfer in the form of cheaper consumer goods and high profits.[3]

World-system theory (WST) was inspired by dependency theory and it offers a 'geographically holistic' perspective on the global economy. Wallerstein (1974) first developed WST as a means for explaining how uneven distributions of economic and political power evolved over time and space. His central thesis was that the contemporary world system, what he labelled the world economy, became global not through imperial

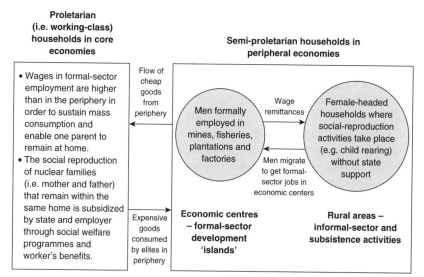

Figure 4.1.1 Unequal exchange and households in the core and periphery
Source: developed from Taylor and Flint, 2000

117

or military power but through the forces of capitalism manifest in surplus-value transfer and unequal exchange between core and periphery (Hall, 2000).

Wallerstein's analysis of these unequal relationships led to the development of a hierarchy of countries with respect to their power in the world-economy system: core countries, peripheral countries and semi-peripheral countries (Hall, 2000). These terms have taken on different meanings since their introduction but today there is consensus about the general characteristics of each type of economy. Core economies (e.g. in Europe, Japan and the USA) are driven by advanced industrial activities and producer services that are supported by powerful state governments, a strong middle class (bourgeoisie) and a large working class (proletariat). Peripheral economies (e.g. in Africa, South Asia, Latin America) are typically driven by natural resource extraction or agricultural commodity production, have weak states, a small middle class and a large pool of low-skill and/or peasant labour. Semi-peripheral economies (e.g. 'emerging' economies such as South Africa, China, India and Brazil) lie somewhere in between core and periphery, having both modern industries and cities while sustaining peripheral attributes such as a large peasantry and extensive informal economies.

The power of WST lies not in this tri-partite categorization but in its explanation for how uneven resources flows between rich and poor countries are sustained and why it is so difficult for peripheral countries to increase their power vis-à-vis core economies. A country's position within the hierarchy of nation-states can improve but only if its politicians and capitalists work together to implement industrial-development strategies that increase productivity, attract capital investments from the outside and create competitive advantages vis-à-vis other states. Achieving success is no easy task as core and peripheral economies are perpetually challenged to balance short-term economic needs with longer-term investments in human capital (i.e. education, training), social welfare, technologies and physical infrastructure (Straussfogel, 1997). Moreover, economic cycles (e.g. Kondradieff waves) play an important role in shaping the core–periphery hierarchy through periodic downturns that challenge existing power structures and which can create opportunities for new 'cores' to emerge. For example, the economic downturn of 2008–2010 has raised interesting questions about whether China will emerge from the semi-periphery to become the next global superpower (e.g. see Arrighi, 2007). Although China's economy has industrialized dramatically in recent years, its ability to do so has been influenced in large part by consumers in core economies (especially the USA), thus demonstrating how a country's ability to improve its position in the global hierarchy greatly depends on the actions in, and the circumstances facing, other countries.

Recent applications of core–periphery

More recently, core–periphery research has emphasized four general lines of inquiry: studies of semi-peripheral economies; critiques of neo-liberal forms of globalization; analyses of the international digital divide; and research on natural-resource peripheries. In the first case, focus is on the unique opportunities and development challenges facing the semi-periphery in the world system. Gwynne et al. (2003) analyze how semi-peripheral countries, what they call emerging economies, are developing within three broad regions: Central, Eastern Europe and Russia; Latin America and the Caribbean; and Asia-Pacific. In doing so, they detail the historical and geographical development of these economies and regions and demonstrate how they represent alternative and unique varieties of capitalism (see 1.3 State).

A second area of core–periphery studies highlights the role of neo-liberal ideology and policy in maintaining and reproducing structural inequality in the world economy. Proponents of neo-liberal policies argue that freer markets, effective private-property rights, reduced industrial regulation and export-led industrialization are the most effective strategies for economic development (World Bank, 1993; Williamson, 2004). Critiques of neo-liberalism are numerous but most critics agree that neo-liberal policies favour advanced industrial regions and sustain the spatial inequality manifest in the global core–periphery divide (Klak, 1998; Harvey, 2006). For example, Arrighi (2002) analyzes Sub-Saharan Africa's post-colonial development history and shows how the 1970s economic crisis led core economies to reshape global economic policies and institutions such that capital flows were redirected to them while African economies were effectively starved of the capital needed to fund development programmes.

119

A third area of interest relates to the role of information technologies (IT) in shaping core–periphery relationships. Building on Castells' (1996, 1998) work on globalization, economic geographers have examined why some regions and communities (e.g. Sub-Saharan Africa, poor inner cities in the USA) are unable to benefit from the new age of information-driven capitalism (see 6.1 Knowledge Economy). Research has shown how digital divides or the uneven distribution of IT and IT-enabled activities creates centres and peripheries with respect to digital information access. For example, Kellerman (2002) describes how particular countries rose to prominence with respect to IT production and use and Gilbert et al. (2008) examine the social digital divide that exists within communities where needy people have limited access to IT (see 2.3 Accessibility).

A fourth area of research strives to understand why resource-abundant regions often remain less developed than resource-scarce ones. A key finding from this work is the notion that a resource curse or addiction can stymie development and chronically marginalize resource-rich regions through structural factors that discourage economic diversification, innovation and social mobilization (e.g. for land rights, labour rights) (Freudenburg, 1992; Auty, 1993). When regions are 'held back' through these factors and forces, they can become resource peripheries that are overly dependent upon core economies' consumption of energy, agricultural products and other raw materials. As Rosser's (2007) study of Indonesia's economic development demonstrates, peripheral status is not inevitable for all resource-rich countries and can be avoided through sound economic policy making and the right external political and economic conditions. In Indonesia's case, its strategic location during the Cold War and its proximity to Japan helped to drive high rates of economic growth during the 1970s and 1980s. All told, resource peripheries play a key role in the world economy and a greater understanding of them can help geographers more fully understand the dynamics and consequences of economic globalization (Hayter et al., 2003).

KEY POINTS

120

- Cores are advanced industrial regions able to sustain comparative advantages over peripheral regions where economies are generally dependent on agriculture, natural resource extraction, and/or labour-intensive and low value-added manufacturing. Semi-peripheral countries (e.g. Brazil, China, India, South Africa) have some characteristics of both core and peripheral economies.
- Dependency theory and world-system theory scholars view core–periphery relationships as the historical products of unequal exchange relationships linking once colonizing economies in the core to their former colonies in the periphery.
- Economic geographers apply the core–periphery concept to studies of emerging economies such as China, India and Brazil, the dynamics and consequences of neo-liberal economic policies, the role of IT in economic development and the challenges facing natural-resource-dependent economies.

FURTHER READING

Makki (2004) analyzes the history of development interventions in the Global South and argues that the current age of 'global neo-liberalism' will continue to reproduce and exacerbate core–periphery inequalities.

See Barton et al. (2007) for an analysis of the unique experiences of the semi-peripheral economies of Chile and New Zealand. Pain (2008) examines the core–periphery relationships linking the city of London and the Southeast England region. Terlouw (2009) details the historical development of the Netherlands and Northwest Germany through a world-systems perspective.

NOTES

1 Rostow's development model posited that all countries – regardless of their history and current material circumstances – could modernize effectively through liberal economic and political policies. He created a development 'ladder' for countries constituted by five stages of growth: 1) traditional society; 2) the pre-conditions for take-off; 3) the take-off; 4) the drive to maturity; and 5) the age of high mass consumption. The model provides a 'one-size-fits-all' non-socialist framework for economic and industrial evolution and in application its promoters argue that developing countries should focus first on exporting agricultural or natural resource commodities and to then direct the profits from these sales toward the establishment of urban based centres of manufacturing.

2 A good example of this is the Firestone (Bridgestone) rubber plantation in Liberia whose 100 million acres of land was acquired from the Liberian government in 1927 for an extremely low price as a result of intense diplomatic pressure from the US government (Fage et al., 1986).

3 Importantly, however, the benefits of this unequal exchange were primarily accessible only to white male workers in core economies. Women and minorities traditionally remained exploited and marginal to the system. These structural inequalities persist today (see 5.2 Gender; 5.3 Institutions).

121

4.2 GLOBALIZATION

Economic globalization is being driven by the geographical dispersal of markets, the functional integration of production activities, and the increasing interconnections and interdependencies between people and places in the world economy. Since the 1980s, popular books about the concept have become commonplace and have generally taken views for (e.g. Friedman, 2005) or against (e.g. Stiglitz, 2002) its contemporary dynamics, meanings and consequences. In academia, analyses of globalization have come from a variety of disciplines including: economics (e.g. Bhagwati, 2004), political science (e.g. Held and McGrew, 2007), sociology (e.g. Sklar, 2002) and cultural studies (e.g. Appadurai, 1996). Economic geographers have demonstrated how circumstances at the local scale, and the spatial connections linking economies, play key roles in determining how places and regions influence and are influenced by globalization (e.g. Kelly, 1999; Bridge, 2002; Dicken, 2004). Because the breadth of the concept is so extensive, some of these contributions are addressed in other chapters of this book (e.g. 1.1 Labour; 6.3 Consumption; 6.1 Knowledge Economy; 6.2 Financialization). Emphasis in this chapter is placed on three themes central to the economic geographies of globalization: the role of multinational corporations (MNCs) in globalization and the New International Division of Labour (NIDL); why there are limits to economic globalization; and how globalization has forced geographers to rethink their conceptualizations of spatial scale.

Multinational corporations and the new international division of labour

As the global economic crisis of the mid-1970s persisted, scholars began observing significant changes in the location of industries and the strategies and structure of large corporations. Hymer (1976) and Fröbel et al. (1978) argued that MNCs were reorganizing the world economy in new ways that differed from the classical (i.e. Ricardian) division of labour; that is, a situation where a country's comparative advantage in trade was based simply upon whether it was a raw material supplier or a producer of manufacturing goods. In this new age of global capitalism, a New International Division of Labour (NIDL) was emerging as manufacturing activities shifted from advanced industrial economies into the developing world. Central to this transformation was MNCs' use of multi-country input sourcing strategies that distributed production across many places and significantly increased the cost-efficiency of manufacturing systems. This spatial reorganization of production was made possible in part by logistics innovations (e.g. containerization) and information technologies (IT) and by changes in the structure of the world financial system (Taylor and Thrift, 1982; Castells, 1996).

The global coordination of production activities has been broadly characterized by three processes: offshoring, outsourcing and original equipment manufacturer (OEM). Offshoring occurs when firms move production facilities to foreign countries in order to reduce production costs while maintaining direct control. Vernon's (1966) product life-cycle theory explained why manufacturing locations for particular products shift from industrial economies to developing countries. This shift occurs when the product leaves the research and development (R&D) intensive stage (which requires high-skilled labour) and when competition for the product shifts from one based in large part on product differentiation (few competitors) to another based on price (many competitors). The need for lower production costs drives MNCs to develop mass-production systems and seek out new factory locations in developing countries where wage rates are lower.

Outsourcing is often used interchangeably with subcontracting. Outsourcing refers to firms externalizing production or services that used to be conducted in-house (e.g. using third-party logistics services instead of owning and operating trucks), whereas subcontracting takes place in a hierarchical organization in which certain tasks are conducted by outside firms (e.g. a small firm supplying nuts and bolts for Toyota). While outsourcing is typically conducted so a firm can focus on its 'core competence', subcontracting can be complex and are driven by the need for speciality parts or services, a desire to achieve cost savings through economies of scope in the subcontracted firm, and/or an intermittent or seasonal need to boost the principal firm's production capacity. Although outsourcing/subcontracting can take place within a country, these terms are increasingly used to refer to MNCs giving contracts to foreign firms to produce goods and services (Dicken, 2007). In contrast to offshoring, which requires more direct capital and

123

managerial investments, international outsourcing/subcontracting can increase the flexibility of an MNC by externalizing some of the costs and risks associated with factory management.

OEM refers to manufacturers with agreements to produce goods to be marketed by other branded distributors or retailers. OEM provides relatively unknown manufacturers access to new markets (e.g. Sears brand white goods – such as washing machines – were manufactured by Japanese firms in the past and more recently by Chinese firms) and may even offer opportunities for technological upgrading, the process of improving product quality, production efficiency and/or the firm's functional capacity through ties to buyers in international markets (see 4.4 Global Value Chains). As Gereffi (1999) demonstrated, OEM has been important for East Asian apparel manufacturers to develop their own brands and clothing design capabilities.

In sum, the offshoring, outsourcing/subcontracting, and OEM have created globally distributed production systems whose effective coordination has been facilitated by innovations in transportation logistics, financial services and IT. This dispersal of production not only makes established MNCs more efficient, it also provides opportunities for newly emerging MNCs (from emerging economies) to internationalize their activities and gain access to new forms of knowledge (e.g. see Tokatli, 2008). However, MNC driven globalization can have negative consequences in both industrialized and emerging economies. In the USA and Europe, globalization of production has increased the wage gaps between blue- and white-collar workers (Bardhan and Howe, 2001). In developing countries, child labour, gender discrimination and sweatshop working conditions have been linked to outsourcing activities in the apparel, textile and other manufacturing industries of the MNCs (e.g. see Rosen, 2002).

The limits to economic globalization

Although MNCs and the NIDL have played an important role in interconnecting and integrating places in the world economy, many business activities remain difficult or impossible to outsource or offshore. Storper's (1992) work on Product Based Technological Learning (PBTL) systems – territorially bounded sets of knowledge and production conventions – showed how regional trade specializations limit the geographical spread of high-value, quality driven production systems (e.g. in software and electronic manufacturing). Because the knowledge needed to be competitive is impossible to internalize completely in a single firm, certain segments of industries high in skill and knowledge intensity (e.g. R&D for IT and biotechnology) can rarely be easily transferred around the global economy. Spatial proximity and industrial clustering are thus rational outcomes of these information and technology needs, and when the right kinds of capabilities evolve in a place or region, convergence or catch-up on the part of other places can be extremely difficult to achieve (Maskell and Malmberg, 1999; Scott, 2006) (see 5.3 Institutions; 3.2 Industrial Clusters; 3.3 Regional Disparity; 2.1 Innovation; 5.5 Networks).

States also limit globalization through their influence on home-based MNCs' investment and trade activities. Dicken (1994) argues that MNCs are first and foremost embedded in the contexts from which they originated and that this embeddedness creates state–firm tensions that limit the geographical reach of corporations (see 5.4 Embeddedness; 1.3 State). In 2002 the 100 leading MNCs still had almost half of all their sales, assets and workers in their home countries (Dicken, 2007). Even when firms are internationalized, the geographical destinations of their foreign investments may be concentrated in particular places or regions. For example, a US firm may appear to be highly globalized when in fact the overwhelming proportion of its international investments are in Canada. Because many states rely on the tax revenues and employment generated by MNCs, governments use incentives (e.g. subsidies, reduced taxes, favourable labour and environmental regulations) to keep firms and jobs local (Yeung, 1998). These conflicts and strategies, when coupled with international differences in state–firm relationships and institutions, make it highly unlikely that the world economy will ever be fully integrated (Jessop, 1999).

Globalization's limits also become clear when one takes a cultural view of place creation and economic development. Amin and Thrift (1993) argue that globalization is in fact creating greater heterogeneity among places which in turn plays a significant role in shaping the world economy. Following Appadurai (1990), they believe that the global economy can be understood in terms of the different knowledge, cultural, people and financial flows that create complex interactions or disjunctures at the local scale. Given the pace and scale of these global–local interactions, cities and regions need to develop effective institutions

125

that enable local actors to access global flows such that they contribute positively to economic development (Amin and Thrift, 1993; Amin and Graham, 1997) (see 5.3 Institutions).

Finally, geographers have also demonstrated that there are significant limits on globalization as a force driving the homogenization of cultures everywhere. While some cultural traits (e.g. Thai food, hip-hop music) have become increasingly globalized in recent decades, ethnic identities, social values and cultural traditions can remain rooted or embedded in places (see 5.1 Culture). Even as international migration connects people and places in new and direct ways, the traditions and identities of migrants are often sustained and reproduced through their diasporic communities (Mohan and Zack-Williams, 2002). Studies have focused on how transnational identities are constructed through international migration (Wong, 2006) and how an immigrant's ties to others within her/his ethnic diaspora create opportunities for transnational investment and trading (Mitchell, 1995). Economic geographers have also examined how diasporic ties contribute to industrial innovation (Saxenian and Hsu, 2001) and how the flow of remittances from immigrants to their home communities constitutes a significant source of financial capital for developing regions (Jones, 1998; Reinert, 2007).

Globalization, states and the rethinking of scale in economic geography

Geographers interested in the political-economic implications of globalization have raised important questions about how it is reshaping the role of the state in regional development. A key debate in this literature centres on whether conceptualizations of geographic scale adequately account for new forms of economic coordination and governance such as those manifest in multilateral and bilateral trade treaties and regional economic agreements. Broadly stated, these debates have sought to move beyond traditional 'containerized' conceptualizations of scale (i.e. local, national, global) by emphasizing one of two general approaches: a relativized view of scale and an anti-scale or topological view.

For proponents of relativised perspectives, traditional conceptualizations of scale fail to account for how economic globalization – driven by neo-liberal capitalism – has forced states to reorganize or rescale their activities in order to carve out competitive niches in the global economy

(Swyngedouw, 1997; Brenner, 2000). Specifically, states have become less concerned with the relationships between their industrial base and the local economy and are instead focused on improving the relationships between local firms and the international marketplace. These rescaling processes are manifest in the creation of new regulatory frameworks that reshape regional growth processes through the development of 'glocalized' (global–local) or 'glurbinized' (global–urban) economic relationships (Swyngedouw, 1992; Jessop, 1999). For example, Grant and Nijman (2004) show how the Indian and Ghanaian governments effectively enabled and promoted uneven development within their regions and cities in order to attract foreign investment. Beyond forcing a reconsideration of how regional development occurs in the context of contemporary globalization, analyses of state rescaling efforts raise important questions about their social justice implications. Specifically, the policies associated with glocalization strategies may exacerbate local inequalities based on class, ethnic identity and/or gender by favouring or prioritizing the needs of external markets and MNCs (e.g. see Nagar et al., 2002).

In contrast, some scholars view scale as an outdated concept and globalization as an outcome of the 'non-scalar' or topological relationships that link places together in the world system (Castells, 1996; Amin, 2002; Marston et al., 2005). In this perspective, firms, industry and regional development are driven by the degree of access to globalized knowledge and capital flows (Hess, 2004). As such, economic, financial and industrial networks, not territorial states, cities or regions, structure the global economy by distributing resources and creating interconnections and interdependencies between places. Although network perspectives have had a significant influence in recent years, most economic geographers recognize that conventional ideas about scale and relativized views of scale (e.g. global–local) remain important for analyses of globalization's causes and consequences (e.g. see Sheppard, 2002; Dicken, 2004) (see 5.5 Networks).

KEY POINTS

- Economic globalization is being driven by the geographical dispersal of markets, the functional integration of production activities, and the increasing interconnections and interdependencies between people and places in the world economy.

- Offshoring, outsourcing and OEM strategies of MNCs play a key role in economic globalization.
- Economic geographers have demonstrated how localized knowledge and capabilities, state power and place-based culture limit the reach of the homogenizing or dispersing forces behind globalization.
- Globalization has sparked important debates about geographical scale's conceptualizations, meanings and manifestations.

FURTHER READING

Wade (2004) provides an empirically rich analysis of whether economic globalization is reducing inequality. Cox (2008) critiques Thomas Friedman's book *The World is Flat* and argues that uneven geographical development is an unavoidable consequence of globalized capitalism. The World Bank's (2009) recent *World Development Report: Reshaping Economic Geography* focuses on the role economic geography plays in shaping countries' abilities to benefit from liberalized trade and investment strategies. See Rigg et al. (2009) for a thorough analysis and critique of the World Bank's arguments.

4.3 CIRCUITS OF CAPITAL

The circuits-of-capital concept was originally articulated by Marx and remains a core idea in radical political economy. In simple terms, the concept helps to explain how economies are constituted by capital-circulation systems that enable capitalists to continually accumulate wealth and power through unequal exchange relationships with workers. For geographers, the idea gained prominence through David Harvey who extended the concept into a critical theory of capitalism and the space economy. In application, it helps to explain how structural factors (i.e. the capital-circulation system) drive uneven development in cities, regions and the global economy. The concept also shows how states and capitalists strive to overcome the crisis tendencies plaguing the capitalist system by altering capital flows through 'spatial fixes'. Geographers have extended the idea into studies of global finance, transnational flows of culture and knowledge and labour migration.

Marx and the circulation of capital

The circuits-of-capital concept, initially developed by Marx in *Capital* (1967), provides a powerful explanation for how capitalism functions. At its core is the notion that capital is circulated within or among places through three primary circuits: the circuit of money capital, the circuit of productive capital and the circuit of commodity capital. These circuits are the primary foundations of the capitalist system and provide the means through which surplus value (i.e. excessive profits derived from unpaid labour) is continually extracted from labour and appropriated by capitalists. Most importantly, the circulation system is a *necessary* condition for the reproduction of capitalism.

Marx's historical-materialist analysis described the evolution of the capitalist system and showed how the development of credit markets, new production technologies and universally recognized forms of money enabled the expansion of capitalist power. Money, a measure of use value, is social power in Marx's theory and the development of money forms that are spatially mobile and widely legitimated is a key innovation in human history. Specifically, in the days of commodity-only trade (i.e. barter) there was a greater balance in the timing and value of exchange that limited, to some degree, the hoarding of wealth and social power. With the development of universal money forms (e.g. coins, dollar bills), exchange relations were transformed from commodity only (C-C) trade to commodity-for-money (C-M) and/or money-for-commodity relationships (M-C). This change creates the conditions necessary for unequal exchange relationships, and one of Marx's key insights was to recognize that while the commodity-money circuit (C-M-C) might achieve a balance between use values of different commodities, with money serving as a place holder of value, the money circuit (M-C-M) could not be sustained if money holders simply gained an equivalent amount of capital after taking on the risk of buying or investing in a commodity and then reconverting that commodity back into an equivalent amount of money. In other words, there needed to be an incentive for taking such risks and this is where capital accumulation through the extraction of surplus value comes into the circulation system. For the system to work, capitalists (the principal holders or 'hoarders' of M) must be able to realize excess profits in the money circuit, represented by Marx as M-C-M' (M' = M + the surplus value extracted from labour). The result was two circuits or systems of circulation, the commodity circuit C-M-C' and the money circuit M-C-M'. Marx then extended these relationships to a third circuit or stage wherein the production of commodities took place (P), although he recognized that this necessary step would temporarily interrupt the flow of capital in the system.

The extraction, exchange and production activities constituted by the money circuit, productive circuit and commodity circuit enable the continual reproduction and expansion of capitalist society (see Figure 4.3.1). Money capital (i.e. capitalist power) is the key driver of the system in that it enables the purchase of the commodities, fixed capital (i.e. the means of production embodied in technologies, machines, a factory, land, etc.) and labour power necessary for circulation. Money is first used for the 'productive consumption' of commodities, the means of production (MP) and labour power (LP) which transforms raw material commodities (C) into manufactured or value-added commodities (C').

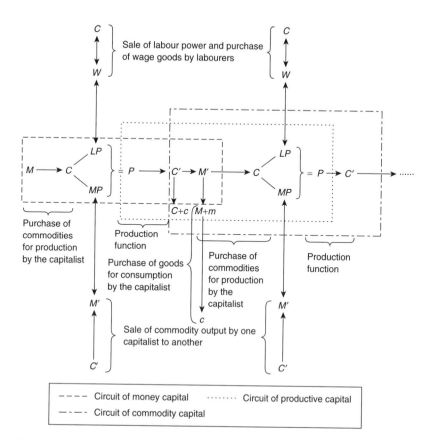

Figure 4.3.1 Marx's capital-circulation system
From M. Desai (1979) *Marxian Economics*. Oxford: Blackwell.
Reprinted with permission.

Labour power receives a wage for its effort, which is then used towards the purchase of wage goods (e.g. basic needs such as food and clothing) that are also produced in commodity circuits. Capitalists, too, consume these goods, but the excess profits (m) they obtain through the exploitation of labour enables them to also purchase higher value or luxury goods (c). Over time, these three stages or circuits continually reproduce themselves in order to expand capitalist accumulation at the expense of labour power and a more equal distribution of material well being in society.

Circuits of capital, uneven development and the spatial fix

Geographers' first significant engagement with Marx's ideas came in the 1970s when David Harvey extended the circuits-of-capital concept to explain economic development processes. Harvey's (1982) landmark book *The Limits to Capital* applied Marx's ideas about accumulation, surplus value extraction and the circulation of capital in the development of a theory to explain how capitalism works and why it needs capital mobility, spatial economic integration and uneven development to sustain itself.

To explain how capital mobility and geography shape the contemporary evolution of an urban, regional or national economy, Harvey (1982, 1989) integrated two additional circuits into Marx's primary ones (i.e. the money, commodity and productive circuits). These parallel circuits are necessitated by capitalism's tendencies towards over-accumulation and devaluation crises (i.e. economic recessions and inflationary crises), and they provide a means for diverting or 'switching' excess profits out of the primary circuits in order to prevent such crises. Moreover, these secondary and tertiary circuits enable states and private-sector actors to continually invest in future means of production and in the social reproduction of labour (see Figure 4.3.2).

132

Secondary circuits – regulated by the state but controlled by private sector actors – provide a key mechanism through which primary capital is diverted into financial markets, fixed capital and the built environment (e.g. roads, stadiums and office buildings). Moreover, this circuit contains a consumption fund to ensure there are sufficient investments in the infrastructure and durable consumer products necessary for commodity consumption (e.g. stoves, shopping malls, refrigerators). Financial and investment markets are central to the secondary circuit and they provide a system through which capital can invest in future expectations for production and profits, what Marx and Harvey call a fictitious form of capital (see 6.2 Financialization).

Tertiary circuits – transfers of excess profits and taxes into science and technology research and social expenditures (e.g. health care, education, security, welfare programmes) – also play a key role in the urban and regional development process in that they provide for public goods (e.g. defence), pay for the costs of governing, fund technological invention and innovation, and facilitate the skilling, training and social

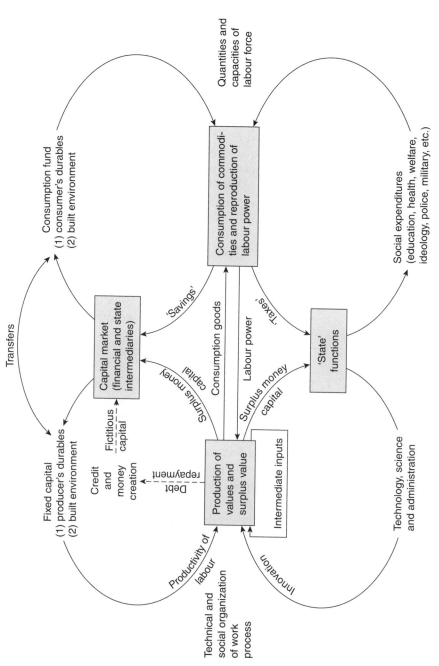

Figure 4.3.2 David Harvey's (1982) circulation of capital framework
Reprinted with permission from D. Harvey (1982) *The Limits to Capital*. London: Verso.

reproduction of labour. States are responsible for determining how much surplus capital and taxes will be channelled away from the primary circuit for these activities. As such, the economic and social welfare policies of a given state (e.g. neo-liberal, developmental, Keynesian welfare state – see 1.3 State) are a central influence on the nature, scale and scope of capital flows in tertiary circuits.

Although Harvey's three-circuits model succinctly describes capital flows within a region, the real-world dynamics of these interrelationships are complicated by the space–time characteristics of capital flows (i.e. from where and when capital is transferred into or from a particular part of the circuitry), shaped significantly by ties to outside markets and places, and determined in part by each region's unique historical and geographical conditions. Moreover, the constant threat of crisis inherent in capitalism means that the circuits need to be continually restructured both internally, through inward transformations that alter the flows within and between the three circuits, and externally, through outward transformations that reshape the city, region or country's ties to other places and circuits. These strategies entail the widening (dispersing trans-locally) and/or deepening (concentrating locally) of capital such that accumulation is expanded and uneven development persists in spite of capitalism's internal contradictions (e.g. production is a social process, but the means of production are privately held; capital strives to be universal while simultaneously benefiting from geographic differences) (Smith, 1990) (see 4.1 Core–Periphery). In other words, capital must continually disrupt the equilibrating tendencies (i.e. the over-accumulation and devaluation process) of the system through 'spatial fixes' that redirect capital flows inward (i.e. locally) or outward (i.e. to other regions and places) (Harvey, 1982; Jessop, 2000; Schoenberger, 2004). For example, Arrighi (2002) believes that the US government resolved the economic crisis of the 1970s through policies that liberalized global financial flows of capital and attracted foreign investors to the United States.

Geographers have examined the spatial fixes used by cities, regions and countries in great detail and demonstrated how they can take many forms and be organized at or between various spatial scales. Within cities, urban-development strategies such as gentrification and the 'new urbanism' offer means for local capitalists to tap into global capital flows (Harvey, 1989; Smith, 2002). At the national scale, states and MNCs have commonly used outward transformations such as the offshoring of manufacturing activities or of shift to export-oriented industrialization policies as a strategy to redirect global capital flows into urban and regional economies (Glassman 2001; Harvey, 2001). Finally, globalization – manifest in the increasing functional integration and supranational regulation of the world economy – has complicated the design, implementation and outcomes of spatial fixes as cities, regions and national economies rescale their regulatory systems in order to make their economies more competitive and receptive to capital flows (Swyngedouw, 1997; Jessop, 2000; Brenner, 2001) (see 4.2 Globalization; 3.4 Post-Fordism). These 'solutions' come at a price, however, in that they may simultaneously facilitate over-accumulation in the global economy (Hung, 2008), create vulnerabilities in the national economy (Glassman, 2007) or reproduce structural inequalities at the urban scale (Bartelt, 1997; Wyly et al., 2004).

Applications and extensions

Beyond contributing to theories on capitalist crises and their spatial 'fixes', analyses of capital circuits have been applied to understand how, where and why particular industries evolve and how the stages of commodity production and consumption are shaped by international capital flows (Foot and Webber, 1990; Christophers, 2006). Capital circuits have also been used to analyze the historical development of slavery and the furthering of racial-segregation practices in the ante-bellum and post-bellum South (McMichael, 1991; Wilson, 2005). Flows of money in the world economy have been informed by the circuits concept, particularly with respect to studies of off-shore banking centres (e.g. the Cayman Islands, Panama) that have attracted large flows of capital through relaxed regulatory frameworks that give them a global competitive advantage (Roberts, 1995; Leyshon and Thrift, 1996; Warf, 2002) (see 6.2 Financialization). Finally, labour migration has been analyzed through a circuits framework with respect to both its contribution to labour supplies in cities and regions and in terms of its support for flows of financial remittances back to home countries (Peterson, 2003; King et al., 2006) (see 1.1 Labour).

Recent work has linked social and cultural concerns to the political-economic foundation of the circuits-of-capital concept (see 5.1 Culture). Hudson (2004) argues that materialist approaches need to be augmented by deeper consideration of how agents, social relations and institutionalized practices create spaces where knowledge, commodities

135

and culture flow and where capital's primary, secondary and tertiary circuits are realized. For Lee (2002, 2006), the circuits-of-capital concept is too limiting in that it focuses solely on economic or capitalist use values at the expense of understanding how value often transcends capitalist ideals and how its diverse meanings are socially contested and geographically and historically embedded (see 5.4 Embeddedness). He suggests that the concept be more inclusive and place-sensitive such that it is better able to account for alternative and/or diverse forms of economic organization (e.g. localized forms of currency, bartering, cooperatives – see Gibson-Graham, 2008).

KEY POINTS

- Marx first conceptualized the circuits of capital as a system that functioned in three stages: the money circuit, the productive circuit and the commodity circuit. David Harvey extended Marx's ideas by considering the spatial aspects of capital circuits and by recognizing the need for alternative circuits (i.e. secondary and tertiary circuits) into which capitalists can divert accumulated capital for investments in infrastructure, industrial development and the social programmes needed to sustain an adequate supply of labour.
- Spatial fixes are strategies employed by capital and states, which strive to prevent or mitigate crises by altering capital circuits through redirecting capital flows inwardly or outwardly.
- Recent extensions of the circuits-of-capital concept can be found in industry studies, financial market analyses, labour/migration research, and in debates on how circuits, flows and values in the world economy are socially constructed and distinctively shaped by places and their histories.

FURTHER READING

Jones and Ward (2004) revisit Harvey's *Limits to Capital* with the goal of extending his theories on the causes of capitalist crises. Bello (2006) examines how global circuits of capital were reorganized in the 1980s and 1990s through the US government's neo-liberal strategies and how they caused the economic and financial crises that doomed the prospects for the full integration of the world economy.

4.4 GLOBAL VALUE CHAINS

The global value chain (GVC) concept explains how the supply chains linking manufacturers and consumers in advanced industrial economies to producers based in developing economies are governed through transnational networks. Beyond mapping global commodity and value flows, GVC also explains why multinational corporations (MNCs) based in advanced industrial economies retain power whereas supplier firms in developing countries experience great difficulties in improving their position in the value-chain hierarchy. Recent studies have focused on how globalized product standards and production certification schemes are shaping value chain structures. Some economic geographers have sought to develop more place- and region-sensitive analyses using the GVC framework.

From global commodity chains to global value chains

The term commodity chain originated within world-system theory (see 4.1 Core–Periphery) to describe how global trade is organized and evolved historically and spatially through periods of expansion and contraction in the world economy (e.g. Schumpeterian cycles, Kondradieff waves) (Hopkins and Wallerstein, 1986). Few commodity chain studies today refer to world-system theory, however, and the concept was transformed into a more industry- and enterprise-centred framework by Gereffi and Korzeniewicz (1994). For Gereffi et al. (1994: 2), global commodity chains (GCCs) are 'networks of labour and production processes whose end result is a finished commodity', and they are constituted as a linear series of nodes linking raw material supply locations to the final markets.[1] Although related, the GCC concept differs from the New International Division of Labour (NIDL, see 4.2 Globalization) in that the NIDL focuses on intra-firm relationships (i.e. within MNCs) while the GCC concept emphasizes the inter-firm networks linking suppliers, producers and buyers in international markets (see 5.5 Networks).

GCCs are constituted by four characteristics: input–output structures, geographical distribution, governance structures and institutional frameworks that create the conditions necessary for their development (Gereffi, 1994; Gereffi, 1995). Input–output structures are the exchange relationships that link raw-material suppliers to producers and consumers. These exchanges result in geographical distributions that are periodically transformed as new lead firms and consumer markets emerge. Governance structures are the rules, regulations and power relations that determine how firms interact in the GCC. These are especially important in influencing if and how lower-tier suppliers can improve their positions in the chain. Institutional frameworks are similar to governance structures in influence but remain distinct in their reference to the wider structural environment, such as institutionalized rules, norms and regulations (e.g. trade policies, product-safety standards, intellectual property right regulations) that enable lead firms to subordinate other firms through their control over access to markets and information (Raikes et al., 2000).

Empirically, GCC research has focused primarily on the governance of commodity chain relationships with an emphasis on the role that lead firms (often MNCs) play in structuring global industries. The main contribution of Gereffi's (1994) early GCC work was his characterization of the apparel industry as a buyer-driven commodity chain, as contrasted against commonly assumed producer-driven commodity chain. Figure 4.4.1 summarizes the management strategies of each type of chain. Producer-driven chains are associated with industries that have high entry barriers due to their need for extensive capital investment and advanced manufacturing technologies (e.g. the automobile industry, advanced electronic equipment). Leading manufacturers (e.g. Toyota) have complete and direct control over the production process and they tightly manage component and parts suppliers. In buyer-driven relationships, which Gereffi argued to be more common, manufacturers play a subordinate role to retailers and distributors, who play the role of the lead firms. Buyer-driven governance is typical for industries characterized by low entry barriers, mature production technologies and lower fixed capital needs (e.g. apparel, toys, agri-food). As a result, production activities in buyer-driven chains are relatively easy to outsource to independent firms or farmers, and power resides not with producers but with retailers and distributors who control branding, design and marketing.

Producer-driven Commodity Chains

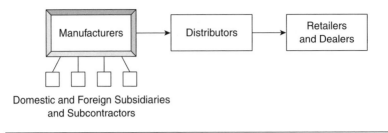

Buyer-driven Commodity Chains

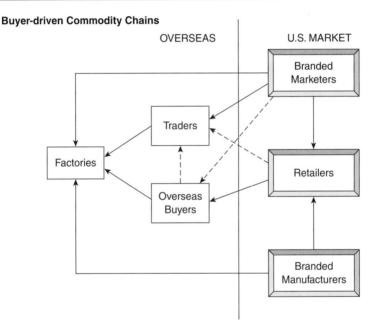

139

Figure 4.4.1 Gereffi's (1994) conceptualization of producer and buyer-driven GCCs
NOTE: Solid arrows are primary relationships; dashed arrows are secondary relationships. Retailers, branded marketers, and traders require full-package supply from overseas factories. Branded manufacturers generally ship parts for overseas assembley and re-export to the manufacturer's home market.

The distinction between producer- and buyer-driven industries is often blurry as lead firms may use a blend of strategies that are associated with both types of governance (Raikes et al., 2000). Recognition of this complexity led to the development of the global value chains (GVC) framework as a means for providing a more subtle distinction between different types of buyer- and producer-driven chains (Humphrey and Schmitz, 2000; 2002). Gereffi et al. (2005) argue that the mode of governance in a GVC is not determined by the commodity *per se* but instead by the nature of supplier–buyer relationships as manifest in the geography of transactions (e.g. are face-to-face meetings essential for contract negotiations?), by the need for high levels of mutual trust and by the technological capabilities of suppliers.

Gereffi et al. (2005) identified five modes of GVC governance (see Figure 4.4.2). Three of these structures are commonly associated with buyer-driven industries (market, modular and relational) and two are more specific to producer-driven industries (captive, hierarchy). At the most basic level are market forms of governance where price is paramount, transactions are arms-length (i.e. low levels of trust), and where the technology needed for participation is relatively standardized and widely accessible. When buyer-driven industries require more intense interactions between lead firms and suppliers, governance takes on a relational or modular form as lead firms and suppliers develop close and mutually dependent relations with a high degree of trust. At the other end of the spectrum are captive and hierarchical GVCs, both of which are producer-driven forms of governance that enable firms to maintain tighter control over production and sourcing. Captive value chains exist where parts and components suppliers are highly dependent on lead firms, whereas hierarchical GVCs refer to those where suppliers are vertically integrated into the production structure of the lead firm. However, as Sturgeon et al. (2008) have observed, these categories of governance are not mutually exclusive to particular firms or industries. Instead, contemporary GVCs are organized through diverse strategies aimed at improving the competitiveness and organizational flexibility.

Recent research of GVC has sought to understand how value chain integration influences local and regional development processes, particularly in the Global South. These studies cover a wide range of industries – automobiles in Brazil and India (Humphrey, 2003), coffee in Indonesia (Neilson, 2008), and food and apparel in Africa (Gibbon; 2003; Ouma, 2010) – and analyze how multinational systems of regulation (e.g. neo-liberal economic reforms, product standards) are influencing the prospects for developing country economies to benefit through

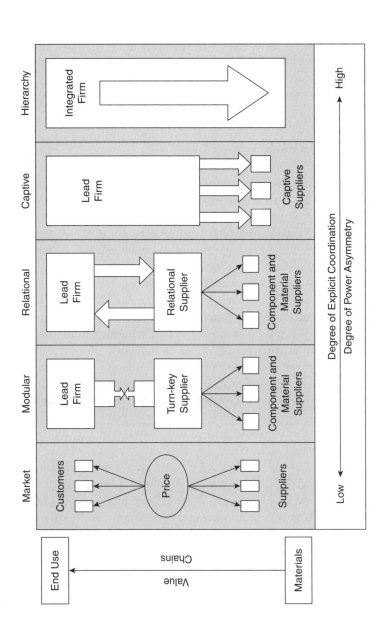

Figure 4.4.2 Five global value chain governance types (from Gereffi et al., 2005).
Copyright © 2005 by G. Gereffi, J. Humphrey and T. Sturgeon (2005) 'The governance of global value chains', *Review of International Political Economy*, 12(1): 78–104. Reproduced with permission of ABC-CLIO, LLC.
Reproduced with permission from Taylor and Francis from STM permissions guidelines. Copyright © Routledge 2005.

ties to GVCs. The findings show how GVC integration is an uncertain, complex and contested process whose outcome is contingent upon the commodity in question, the governance strategies of lead firms, and on the ability of states and workers in developing countries to gain concessions with respect to working conditions, wages, and the value-added they contribute to the GVC (Gibbon and Ponte, 2005).

Upgrading in GVCs

GVC research also focuses on how value chain relationships can enable suppliers to improve their technological capabilities. Changes in technological capabilities of suppliers are conceptualized as industrial upgrading and take a variety of forms: 1) improvements in the design or quality of products; 2) increases in production efficiency and productivity; 3) changes in their role in the value chain (e.g. a parts manufacturer becomes a parts designer); or 4) diversification to new sectors or industries (Gereffi, 1999; Humphrey and Schmitz, 2002). Studies of industrial upgrading analyze how suppliers succeed or fail to improve technological capabilities and positionality in GVCs through industrial upgrading in order to capture more of the value-added. Beyond the need to develop appropriate skills, efficiencies and infrastructures, the prospects for upgrading also hinge upon forms of governance (Gereffi et al., 2005). In some cases, lead firms may prevent upgrading through technological advances that deskill or devalue the supplier's labour (e.g. technologies that reduce the importance of raw-material quality in the cocoa or cotton-processing industries). In other cases, safety, sanitary, environmental and/or labour regulations create barriers for upgrading (e.g. in horticultural-processing industries). Moreover, pricing expectations reward only those suppliers able to compete in high-volume and low value-added markets (e.g. in apparel manufacturing) and therefore may discourage upgrading (Fold, 2002; Barrientos et al., 2003; Gibbon and Ponte, 2005).

142

Studies of industrial upgrading have also shown how links between GVCs and clusters can improve the production capabilities, competitiveness and resilience of local industries and technical communities (Humphrey and Schmitz, 2000) (see 3.2 Industrial clusters). For example, Giuliani et al. (2005) show how global buyers who source from clustered industries (e.g. shoemaking, food processing) often encourage suppliers to upgrade their process and product capabilities such that they meet cost and quality standards. Importantly, however, clusters alone will not enable upgrading (e.g. see Izushi, 1997; Schmitz, 1999). States, industrial-promotion agencies and businesspeople must help to create the appropriate institutional conditions for the successful upgrading of clustered firms (Okada, 2004) (see 5.3 Institutions).

Recent debates and extensions

Economic geographers have extended the GVC concept in two directions. The first strives to understand how new regulatory and product certification systems (e.g. ISO 9000 and 14001) are influencing GVC relationships. Particularly significant are studies that assess the role of regulatory structures in creating GVC entry barriers for firms in developing and emerging economies (e.g. Neumayer and Perkins, 2005) and those that analyze the effectiveness of product certification programs (e.g. FSC certification for forest products) in contributing to fair trade and sustainable development (see 6.3 Consumption; 6.4 Sustainable Development). Although suppliers and workers who commit to these regulatory frameworks expect better prices and working conditions, these are in practice difficult to achieve, as monitoring is often costly and implementation is highly problematic (e.g. see Morris and Dunne, 2004; Hilson, 2008).

143

A second area is the development of the global production networks (GPN) framework. The GPN concept was developed by economic geographers dissatisfied with the GVC emphasis on linear, unidirectional and vertical relationships linking lead firms to suppliers and, ultimately, the consumers (Dicken et al., 2001; Henderson et al., 2002). Unlike GVC approaches that emphasize sectoral development, GPN studies take a territorial view in understanding regional development processes, and stress the role of region-specific characteristics in embedding production activities (Coe et al., 2004)(see Embeddedness). The distinguishing feature of GPN is in its emphasis, which not only incorporates inter-firm alliances, but also public–private partnerships, such as relationships between firms and civil-society organizations such as environmental protection groups and labour unions. These extra-firm relations are both local and trans-local, and often introduce additional complexities that blur 'traditional organizational boundaries' (Henderson et al., 2002: 445).

Empirical applications of the GPN framework are relatively new, with debates continuing as to whether the concept can offer significant improvements over GVC approaches (Coe et al., 2008; Sturgeon et al., 2008).

KEY POINTS

- The GVC concept is used to examine how suppliers based in emerging and developing regions are linked to leading MNCs that manufacture products (e.g. automobiles) for or sell retail goods to consumers in advanced industrial economies.
- Research on GVCs largely focuses on their governance – i.e. the rules, institutions and conventions that regulate value chain relationships. Studies also exist on the prospects for suppliers, typically based in developing economies to upgrade their technological capabilities and capture more value-added in the GVC.
- Recent studies analyzed how changing regulatory systems (e.g. product certification) are reshaping GVC. Also, the global production network (GPN) concept was developed as an extension of GVC analyses and it emphasizes place and region-specific factors as influences on global market integration processes.

144

FURTHER READING

The Global Value Chains Initiative website (http://www.globalvaluechains. org) provides a useful and highly accessible set of resources about GVCs. Riisgaard (2009) examines the East African cut-flower industry and details how the use of private social standards for working conditions might improve the quality of the relationships between labour and business owners. Hudson (2008) explores the global production networks literature and assesses how it might better engage with studies of cultural political economy.

NOTE

1 The GCC concept is related to the French *filière* method for studying commodity flows but is different in that it offers a more coherent and unified theoretical approach for studying production and consumption networks (see Raikes et al. (2000) for details).

Section 5
Socio-Cultural Contexts of Economic Change

How do economic geographers conceptualize the 'economy' in contemporary settings? Although the economy has been at the core of analysis for economic geographers, they also increasingly acknowledge that territorial dynamics are multi-dimensional and that economic processes are deeply intertwined with social processes, cultural transformation, and institutional change. Section 5 includes key concepts that not only serve as key geographic determinants, but also represent analytical tools to understand socio-cultural contexts of the economy.

In the chapter on Culture, we examine how economic geographers engage with cultural aspects of economic change. If the chapter on Regional Disparity (Section 3) represented the central ideological debate of the past, culture represents the contemporary realm in which major ideological differences are emerging. We review the renewed importance of culture prompted by the inter-disciplinary 'cultural turn', as well as the characteristics and critiques of post-modernism and post-structuralism; we also describe recent culturally oriented research, including studies on global convergence, conventions and norms, and the cultural economy. You will note not only the irreconcilable paradigms co-existing in the discipline today, but also the diversity in the themes engaged by economic geographers in the name of culture.

Under Gender, we explore the contributions of feminist economic geography in better understanding variations in access to economic opportunity and in the economic processes that create regional diversity. In this chapter our focus is on how attention to gender, particularly as it relates to labour-market processes, has improved explanation in economic geography. Throughout this chapter, we show how gender is strongly implicated in economic-geographic processes, such as economic restructuring and globalization, and how such processes, in turn, shape gender.

Institutions play a central role in shaping economic geographies by providing structural forms (e.g. organizations, rules, behaviour patterns and norms) that guide economic actors and coordinate their activities. While the historical development of global capitalism has been driven in large part by the evolution of common institutional forms such as markets, firms and property rights, it is also important to recognize that the social, cultural and political characteristics of these structures vary geographically. Institutional differences can make it difficult for firms in different places to become integrated through trade and investment relationships, while, at the same time, such differences play a central role in creating competitive advantages (or disadvantages) for regions. Beyond examining the role of institutions in regional development and globalization, economic geographers have also demonstrated how institutions can create and/or reproduce regional and social inequalities such as those associated with gender or racial discrimination.

In the Embeddedness chapter, we show the important role of personal and social networks in territorialization of economic activities. Drawing initially on the work of Karl Polanyi and Mark Granovetter, economic geographers have studied embeddedness and its significance from three primary perspectives. First, national and regional economies are viewed as being embedded in distinct cultures and political

146

economies, which create varieties-of-capitalism in the contemporary world economy. Second, the embeddedness or rootedness of economic activities in industrial and social networks has greatly improved our understanding of clusters, learning and knowledge-transfer processes, and livelihood strategies. Third, regional stagnation can be explained in part by an economy's embeddedness in place-specific institutions that may prevent innovation or progressive industrial change by 'locking-in' particular industrial sectors or ways of thinking about economic development. In addition to these constructive applications of the concept, economic geographers have also debated the utility, scalability and applicability of the embeddedness concept.

Section 5 concludes with a chapter on Networks, which represent something of a new approach to understanding the economy. Networks are socio-economic structures that connect economic actors and facilitate flows of capital, information and commodities. Since the 1990s, networks – as forms of economic and social organization – have been central objects of analysis in economic geography and economic sociology. Economically, networks help to sustain agglomeration economies and clusters, facilitate information-driven capitalism, and organize the world economy through a hierarchy of cities wherein financial markets, producer services, transportation systems and corporate functions are controlled. Socially, inter-personal relationships, mutual trust and social capital based on racial, class or gender (e.g. 'old-boy' networks, ethnic networks) have been acknowledged to play a role in shaping economic activities. More recently, economic geographers have drawn on actor-network theory to analyze the micro-social processes, and in particular, those that involve non-human agents (such as the environment).

5.1 CULTURE

Culture in essence is a system of knowledge that includes both codified and tacit dimensions. Culture may be associated with a place ('regional culture') or with an economic agent ('corporate culture'). To contemporary economic geographers, culture plays a multi-dimensional role in economies; it is a resource, an endowment, a factor input, an intervening variable, as well as a product and an outcome. Culture is a source of what makes a place unique and what connects a place to other places (i.e. networks). Culture is an organizing principle and a reference for decision making.

The 'cultural turn', according to Thrift (2000a), occurred within various disciplines that were 'wrestling with the problem of how to study the economic as a cultural formation' (p. 689). In economic geography, the 'cultural turn' refers to a paradigmatic shift that took place in the late 1980s as a result of a growing interest in culture and cultural studies. This shift had two major components, epistemological and thematic. The epistemological shift followed the already on-going reorientation of economic geography at the time, from quantitative methods and quasi-scientific approaches to qualitative methods and humanistic approaches; it actively incorporated influences from cultural studies and critical/social theory (post-modernism and post-structuralism) as well as debates on subjectivities and identity from feminism and cultural geography. In contrast, the thematic shift highlighted the persistent significance of cultural variations as foundations of industrial competitiveness, institutional effectiveness and regional growth, and placed renewed emphasize on the role of cultural factors in economic-locational decisions. Contemporary research on culture is linked to various other traditions and is represented by three major trends: the cultural dimensions of economic globalization, cultures as institutions and their role in economic development, and cultural industries and the production of culture.

The 'cultural turn': epistemological debates

Critical theorists such as Foucault's systems of knowledge or 'discourse' analysis (1981) and Derrida's (1967) deconstructionism influenced geographers' work during the cultural turn (see, for example, Yapa, 1998;

Barnes, 2001). In particular, epistemological debates in economic geography with respect to culture invited influences from post-modernism and post-structuralism which became influential in cultural studies and critical/social theory in the 1970s, and from feminism and cultural geography.

Post-modernists attacked theories that were considered part of the 'modern', i.e. scientific, rational and mechanistic (see Dear, 1988). Post-modernism was also in part a critique of structuralism, including that of Marx and Freud. Harvey (1989) portrays post-modernism as 'a rejection of "meta-narratives (large-scale theoretical interpretations purportedly of universal applications)"' (p. 9). Soja (1989) embraced post-modernism and developed a geographic paradigm by adapting and spatializing critical social theory, which he called 'spatial hermeneutic', in order to establish a more 'interpretive human geography'.

Post-structuralism emerged as a segment of post-modernism and stressed 'a theoretical approach to knowledge and society that embraces the ultimate undecidability of meaning, the constitutive power of discourse, and the political effectivity of theory and research' (Graham-Gibson, 2000: 95). Post-structuralism rejects the assumption of an evolutionary succession of knowledge, and instead seeks to restructure epistemological categories established under modernism. Such attempts are made with the goal of revealing implicit biases in epistemological categories and boundaries of knowledge, both of which are viewed as manifestations of social power relations that function largely to reinforce the status-quo (Dixon and Jones, 1996). Post-structuralism, however, did not receive widespread recognition by the segment of economic geography that continued to cater to 'scientific geography' (Dixon and Jones, 1996) and its positivistic assumptions. What was observed instead was the traditions of phenomenology and social psychology that influenced a segment of geographers through actor network theory (see 5.5 Networks).

Feminism has also been influential by revealing the role of identity and subjectivity in the formation of multiple realities (see, for example, Butler, 1990; McDowell and Court, 1994; Pratt, 1999) and pointing to the danger of class reductionism[1] (Ray and Sayer, 1999) (see 5.2 Gender). At the same time, cultural geographers began questioning the notion of culture in the 1990s, with some arguing against viewing culture as ontologically given and advocating instead the study of the ideology of culture (Mitchell, 1995; Castree, 2004). These theoretical developments contributed the view that the reality is both multi-faceted and multiple, effectively calling into question the long tradition of positivism which had been a dominant and implicitly accepted paradigm in economic geography (see Dear, 1988; Dixon and Jones, 1996).

The very characteristics of post-modernism and post-structuralism, which are 'profoundly destabilising and potentially anarchic' (Dear, 1988: 266) turned out to be its major weakness. As Harvey (1989) noted, post-modernism 'swims, even wallows, in the fragmentary and the chaotic currents of change as if that is all there is.' (44) Martin (2001) was concerned with the tendency of the 'cultural turn' that emphasized, in his words, '"sexy" philosophical, linguistic, and theoretical approaches' (189). Markusen (1999) warned geographers not to encourage students to pursue 'fuzzy concepts' or concepts that lack substantive clarity, and Martin and Sunley (2001) cautioned against 'vague theories and thin empirics'. Ray and Sayer's (1999) damning characterization of the cultural turn as being 'largely endogenous to academia' (p. 2) suggested that, at least to some, the cultural turn only distracted scholars from many other important disciplinary objectives. Martin (2001), for example, was particularly concerned with the declining relevance of research in economic geography to policy makers. Ray and Sayer (1999) also lamented the retreat of feminism from economic aspects of gender, and found the tendency particularly perplexing as women suffer more from economic problems than men.

151

Looking back from the vantage point of 2010, in spite of its stated objective, the 'cultural turn' was perhaps as hegemonic a fad as the French Regulation School (see 3.4 Post-Fordism) of the previous era in its influence and language. Although the 'cultural turn' has led many economic geographers to take post-positivism[2] seriously, the group of scholars who catered to this view largely disbanded to pursue their own individual topics, encouraging further fragmentation of the discipline.

The 'cultural turn': thematic debates

Within economic geography, culture has long been present, if only implicitly. Economic geographers' engagement with culture goes back to Marshall's (1920[1890]) observation of 'industrial atmosphere' (see 3.2 Industrial Clusters). Culture has long been recognized as important to processes that lead to spatial variations. Debates on industrial organization that dominated economic geography starting the late 1970s, for example, recognized various alternative models to the American scientific management. The Regulation School and debates on flexible specialization acknowledged cultural underpinnings of the industrial districts in the Third Italy, or of lean production practised by Toyota Motors in Japan, had an important role in sustaining the competitiveness of regions and industries (see 3.4 Post-Fordism). More recently, Schoenberger (1997) and Thrift (2000b) studied how corporate culture and the culture of managers inside the firm. Ettlinger (2003) explored practices in workplaces that involve multiple rationalities and different types of trusts to illustrate the relational and microscopic nature of social interactions.

The 'cultural turn' bestowed legitimacy on research that deals explicitly with culture and economy. In particular, influenced by interdisciplinary developments, economic geographers began questioning the positivistic assumptions of the 'economic' (Thrift and Olds, 1996; Thrift, 2000a; O'Neill and Gibson-Graham, 1999). According to Thrift (2000a), the economy became increasingly viewed as a discursive phenomenon, or a form of rhetoric, in the 1990s (Thrift, 2000a: 690–1). Thus, the 'cultural turn' in economic geography rested on the view that cultural and economic processes were mutually constitutive, a view that, as we shall show in the subsequent section, has generated several research trajectories (Thrift, 2000a).

152

In subsequent sections, we discuss three emerging research fields in economic geography that intersect with culture: 1) cultural dimensions of economic globalization, including cultural convergence, divergence, or persistent heterogeneity; 2) cultures and the role of informal institutions, such as norms and practices, expressed as corporate or regional cultures; and 3) cultural economies, the production of culture, and cultural/creative industries.

Culture under globalization

Globalization has been equated with the global convergence of culture, including the standardization of business norms, practices and mindsets (see 4.2 Globalization). Economic geographers dispute such views, pointing instead to the persistence of place-to-place differences across the globe. Rutherford (2004) took the case of lean production in Canada, the United Kingdom and Germany and analyzed how this production method, which originated in Japan and was adopted globally, does not necessarily lead to convergence in workplace regimes but instead may entail the possibilities of increased divergence across nations. Christopherson (2002) explored persistent differences in national economic practices, especially with respect to labour market practices.

International migration is a key driver of cultural diffusion and transfer. As acknowledged as one important aspect of Appadurai's (1996) kaleidoscopic interpretation of globalization,[3] contemporary international migration not only introduces new ethnic cuisines but also shapes distinctive urban landscapes and sustains entrepreneurial culture. K. Mitchell (1995), for example, discussed the role of Hong Kong immigrants in altering landscapes in the city of Vancouver, Canada. Saxenian (2006) illustrated how the global diffusion of start-up culture was in part facilitated by the development of risk capital in various parts of the world, combined with the global mobility of entrepreneurs. In the past, the mobility of highly educated people from the developing to the developed world (the 'brain drain') was considered detrimental to the future of developing countries. Saxenian suggested that 'brain circulation' represents the new trend in global mobility. Taking the case of Silicon Valley, she argued that entrepreneurs who originally came from Taiwan, China and India to attend graduate schools in the United States to earn advanced degrees in science and engineering played a major role in linking Silicon Valley to these emerging economies by injecting dynamism in the Valley as well as transferring scientific knowledge and business know-how to the emerging economies.

153

Another emerging area of study under culture and globalization is tourism, an example of how production and consumption are intricately connected through global–local links. Considerable diversity is observed in tourism today, from conventional sun-and-beach tourism, to cultural tourism (which includes heritage tourism) and eco-tourism. Cultural tourism, a form of contemporary pilgrimage, is motivated by the belief that only certain places offer authenticity in culture, arts and music. For example, Elvis Presley's mansion Graceland in Memphis, Beatles-related locations in Liverpool and the Bob Marley Museum in Jamaica are all results of successful place-marketing aimed at taking advantage of the search for authenticity (Connell and Gibson, 2003).

Tourists' influences on culture are often described as cultural imperialism, but demand generated by tourism can enable a regional art complex to prosper. For example, in Bali, Indonesia, music and dance developed not in isolation but in conjunction with and as a result of international tourism. In Southern Spain, flamenco music and dance evolved along with the rise of tourism, with the state serving as a co-producer of tourism (Aoyama, 2007, 2009). Tourism inherently involves the state which governs regulatory frameworks, infrastructure development and local-development planning. In Singapore, the state played a major role in re-valorizing local cultural assets to cater to demand from tourism (Chang and Yeoh, 1999). Finally, the growth of tourism has implications for the conditions of work and protection of worker rights. Terry (2009), for example, examined the changing legal status of Filipino seafarers in the cruise ship industry, showing how the cruise industry has created an international legal space that increasingly jeopardizes the human rights of cruise-ship workers.

Culture as institutions

Culture is notoriously difficult to operationalize and measure quantitatively. While economists typically have used indicators such as trust as measures of culture and have explained human behaviour through game-theory, economic geographers generally claim that the analytical usefulness of culture is severely limited when the concept is quantified. Yet systematic analysis of culture remains challenging. To operationalize culture, some economic geographers opt to analyse institutions in order to understand the role of socio-cultural contexts in economic change, because the cultural and social underpinnings of economic activities function as informal institutions that affect the behaviours of economic actors (see 5.3 Institutions). Storper (2000), for example, arguing that culture lacks specificity as an analytical tool, prefers to focus on conventions that signify 'the simultaneous presence of three dimensions: the rules of spontaneous individual action; constructing agreements between persons; and institutions in situations of collective action' (p. 87).

The varieties-of-capitalism (VOC) thesis developed by Hall and Soskice (2001) acknowledged that institutional similarities and differences across national economies arise from sanctions and incentives that affect economic behaviour (see 1.3 State). Strategic interactions among firms, types of market economies, forms of institutions and organizations, culture, rules and history all contribute to the formation of different forms of capitalism. Hall and Soskice are particularly interested in what they called the 'institutional comparative advantage' of a country, which has implications for policy making. Gertler (2004) examined the cultural underpinnings of industrial practices and firm behaviours among advanced machinery users in Germany and North America, showing the complexity of tacit knowledge production and transfer.

At the regional scale, research on institutions typically examines the effectiveness of formal institutions, such as local and regional governmental agencies, regional consortiums, voluntary associations, laws and regulations (see Lawton-Smith, 2003). However, the core of regional culture, such as social norms, collective mindset and responses to incentives, remains tacit and intangible (see 5.4 Embeddedness). These informal dimensions of regional institutions, according to North (1990), tend to persist long after formal institutions have been altered. Saxenian (1994) compared Silicon Valley and Boston's Route 128 and showed how differences in the two regions' cultures have influenced entrepreneurs' risk-taking behaviour, allowing Silicon Valley to grow, in spite of comparable formal institutions such as universities and lead firms in the two regions (see 2.2 Entrepreneurship).

155

Cultural industries and production of culture

While art and culture have been subjects of economic inquiry since the 1960s, economic geographers began acknowledging the importance of cultural industries as a source of employment in major metropolitan areas only in the late 1990s (Pratt, 1997; Scott, 1997; Power, 2002). Cultural industries – comprising the learning, display and sale of literary and visual arts, crafts and music – are grounded in uniquely place-specific cultural heritages. Skills in cultural industries are often based in tacit and uncodified knowledge, and as a result, proximity and agglomeration still matter for cultural industries. Research on cultural industries includes a variety of themes, such as the significance of cultural workers (Christopherson, 2004) and the role of amenities in urban growth (Florida 2002; Markusen and Schrock, 2006). Some take an industrial-district approach, emphasizing the urban dimensions of cultural industries such as the film industry in Hollywood (Scott, 2005) and Vancouver (Coe, 2001), the jewellery quarter in Birmingham (Pollard, 2004), and design occupations in Toronto and Montreal (Leslie and Rantisi, 2006; Vinodrai, 2006). Others have taken a commodity-chain approach to understand the furniture industry (Leslie and Reimer, 2003) or an evolutionary approach in studying fashion and video-game industries (Rantisi, 2004; Izushi and Aoyama, 2006).

These studies have expanded the scope of research, on what is now known as the 'cultural economy'. Amin and Thrift's (2004) *Cultural Economy Reader* provides an inter-disciplinary overview of this work by economic and cultural geographers as well as sociologists; the book includes themes ranging from production and consumption to the 'economy of passions'. Indeed, research on cultural industries and the role of creativity in economic change intersects most closely with cultural geography and sociology (see 6.3 Consumption). Cultural geographers' work, such as Crang's (1994) on 'performativity' taking the case of restaurants in London, and Gregson and Crewe's (2003) study on second-hand shops in the United Kingdom, prompted economic geographers to analyse new industries that closely follow contemporary cultural trends. Molotch (2002) offered a unique angle on the relationship between products and place, with a focus on industrial design.

Furthermore, taking cues from such sociologists as Bourdieu (1984), Latour (1987) and Urry (1995), attention to culture has paved the way for the emergence of consumption studies in economic geography, with some focusing on the convergence of the economic with the cultural (Jackson, 2002) while others have considered the commodification of culture (Sayer, 2003). The rise of cosmopolitan consumerism, i.e. the demand for distinctive cultural experiences particularly in advanced industrialized societies, contributes to the expanding market for cultural commodities. In *Cities of Civilization,* Hall (1998) refers to the importance of the affluent, new generation of consumers which facilitated the diffusion of Chicago blues in the 1950s. Music is perhaps the most immediate cultural product to result from the geographic mobility of people. Consumers seeking leisure and entertainment in the developed world are increasingly oriented toward discovering a unique, distinctive and sometimes personalized 'experience' in return for their time and money (see 6.3 Consumption).

The diffusion and popularity of music today is also facilitated by technology, from the radio to the Internet. The popularity of indie rock (independent rock'n'roll bands) among British and American youth, world music which repackaged various non-western music into new products, or the popularity of Korean actors formerly completely unknown among the Japanese audience are examples illustrating the popularity of art forms considered 'foreign', 'exotic', 'indigenous', 'unspoiled' or 'distinctive'. It is clear that contemporary cultural change is not a unilateral process of global cultures invading local cultures, with local cultures resorting to resistance and preservation as the only means of survival.

KEY POINTS

- The 'cultural turn' was an inter-disciplinary phenomenon, and involved the strong influence of post-modernism, post-structuralism, feminism and cultural studies in various social-science disciplines.
- Although the 'cultural turn' contributed to legitimizing the study of culture in economic geography, it was also heavily criticized for fuzzy concepts, vague theories and thin empirics.
- Economic geographers today engage with culture in a multi-dimensional manner: some emphasize cultural dimensions of globalization, while others prefer to engage with conventions and norms that are created and transferred at various geographic scales. In addition, the role of institutions in economic growth represents one research trajectory, and research on cultural industries, known as the cultural-economy approach, is another.

157

FURTHER READING

For a recent overview of the cultural-economy perspective, see Amin and Thrift (2007). For a critique of the cultural-economy research, see Gibson and Kong (2005).

NOTES

1 The tendency in some Marxist literatures to treat class as overly deterministic, with every explanation reduced to class differences.
2 Post-positivism refers to a belief that positivist views essentially hold validity while acknowledging their inherent subjectivity. Unlike positivists, post-positivists are open to alternative interpretations of what constitutes scientific process.
3 Appadurai (1996) conceptualized globalization as five distinctive 'scapes': ethnoscapes, technoscapes, finanscapes, mediascapes and ideoscapes.

5.2 GENDER

Gender refers to perceived differences between women and men and to the unequal power relations that are based in those perceived differences (Scott, 1986). Ideas about, and practices of, gender are created through everyday life and also help to structure everyday activities as well as broader social and economic processes. As a result, specific meanings and practices of gender vary from place to place and play an important role in creating economic geographies (McDowell, 1999). This chapter explores how gender is implicated in economic-geographic processes, such as economic restructuring and globalization, and how such processes, in turn, shape gender. After examining how concepts of gender have changed in economic geography, we examine specific examples of how attention to gender, particularly as it relates to labour-market processes, has improved explanation in economic geography.

Changing concepts of gender in economic geography

Until relatively recently, economic geography ignored gender as well as other dimensions of identity, such as ethnicity, race, class or sexuality, that influence economic-geographic processes. During the quantitative revolution of the 1950s and 1960s, the key actor animating models in economic geography was economic man, that improbable figure who possesses complete information and always behaves in a rational, utility-maximizing manner (see 1.2 Firm). An important trademark of economic man is, therefore, that neither life experiences nor identity affect decision making.

With the advent of behavioural geography in the 1970s came the acknowledgement that not all decision makers are identical: they have varying amounts of information and diverse backgrounds, goals and constraints, all of which affect their decision making. This recognition of diversity was the beginning of economic geographers' admitting the salience of difference. It is not insignificant that most geographers who have been asking questions about the role of gender have been women, and, in fact, gender has gained attention in economic geography as the number of women in the field has increased. In other words, the life experiences of scholars do influence what they see as important and their explanatory frameworks.[1] In addition to seeking to enrich economic-geography theory by incorporating gender, feminist economic geographers, through their work, seek to improve human well-being, especially that of women and children. Some scholars incorporate gender in their research frameworks without pursuing the social-justice goals that feminists explicitly embrace.[2]

Just as economic geographers did not 'see' gender until fairly recently, the ways in which scholars have looked at gender have shifted over time. In general, the shift has entailed a move away from seeing gender as being rooted in a biological, universal, and unchanging male/female binary to seeing gender as being socially constructed and therefore highly contextualized and mutable; in this latter view, specific meanings and practices of gender depend on time, place and other axes of difference such as class, race, ethnicity, age and sexuality (see, for example, McDowell, 1993). Despite their theoretical incompatibility (in that one view considers gender as naturalized and fixed, whereas the other sees it as unfixed and dependent on context), each of these two ways of seeing gender still retains potency, and therefore both still need to be recognized and held in tension with each other in studies of gender.

159

Early studies introduced the then radical notion that gender matters, for example in spatial behaviour, by exploring gender differences, showing that the daily activity patterns of women and men were quite different owing to distinct gender roles (e.g. Hanson and Hanson, 1980; Tivers, 1985). Although early work on gender did recognize some differences among women (e.g. married with no children; married with children; single), the emphasis was on gender difference, and authors did not problematize the category woman or explore the relationship between diversity among women and gender politics.

The male–female binary view of gender has long been associated with the division between public and private spheres and places, in which the private (home and residential neighbourhood) sphere is considered the woman's realm, whereas the public (workplace and city centre, for example) sphere is a male domain. As economic activity was associated solely with the public sphere, women's domestic work, which was not considered part of the economy, was invisible in economic geography. One contribution of feminist economic geographers has been to open up the categories of public and private and reveal the heterogeneity within each one as well as the myriad connections and interdependencies between the two (Hanson and Pratt, 1988).

As views of gender have changed, recent studies treat gender as just one of many sources of difference among people, that is, as one of the many dimensions (and not always the most salient one) of a person's ever-changing identity and subjectivity. Pratt's (2004) study of Filipina nannies in Vancouver, Canada, for example, demonstrates the importance of the nannies' tenuous immigration status to the economic geographies they live and create. The recognition that gender intersects with other dimensions of identity problematizes the very categories *woman* and *man*, in that differences within each of these categories can be as important as – or more important than – a shared gender identity. Although the nannies and their female employers in Pratt's study were all women, their political interests were substantially different.

Pratt's study illustrates another important shift in geographers' conceptualization of gender, namely the increasing recognition that discourses[3] at varying geographic scales affect the meanings and practices of gender. Pratt (1999) shows how discourses about the Filipina nannies in Vancouver effectively work to position them as marginalized workers and to ensure that their occupational mobility is extremely limited. Similarly, Wright's (1997) study of Mexican women working in *maquiladoras* demonstrates the power of discourse to create and reinforce distinctive categories of workers based on their gender, nationality and ethnicity; Wright also shows that to advance in the workplace Mexican women must – and can – consciously subvert the discourses that have defined them as docile, unskilled labour. In addition to demonstrating the significance of discourse, these studies illustrate the shift in economic geography studies of gender from a focus on power relations at individual and household levels to seeing the importance of gendered structures of power at other spatial scales.

Changing conceptions of gender have prompted reflections on epistemology and methodology. Feminist researchers have increasingly moved towards thoughtful engagement with research subjects, viewing knowledge creation as a joint project with research participants rather than the result of unproblematically tapping or mining the experiences of others (Gibson-Graham, 1994) (see 5.1 Culture). In addition, feminist economic geographers have shifted towards more self-conscious engagement with the subjects of their research to effect positive change. For example, Nagar (2000) describes the complex and often contradictory process of working with a women's group in India to enhance their lives and livelihoods.

160

The fact that everyone has multi-faceted identities that change over time and place points to the importance of geography in the construction of gender, as well as to the importance of gender to economic geographic processes. The next section provides some examples of how gender has changed understandings of economic geographic processes.

Gender and work

Economic geography's longstanding neglect of gender had the serious consequence that many key processes shaping the contemporary world were simply not recognized. Because gender so thoroughly infuses the world of work – paid and unpaid – most of these poorly understood key processes have had to do with the gendering of work, which has shaped economic geographies around the world. In fact, women's work became more visible to economic geographers when women joined the paid work force in large numbers and thereby entered the officially sanctioned realm of the economic (see 1.1 Labour).

Women did not enter the work force as men's equals, however, as one of the most enduring aspects of the labour market has been the separation of women and men into different jobs. Although not all women bear and raise children, and although – for those who do – child-bearing and child-rearing are limited to a relatively small proportion of women's adult lives, and although some men carry heavy child-rearing responsibilities, the strong association of reproductive activity (including care of household, family and community members other than children) with being female has endured, with major repercussions for women's productive activity. The primary vehicle for these repercussions has been the segregation of women and men in distinctly different lines of work, whether that work is agricultural, industrial or commercial and whether it takes place in industrialized or developing countries. In 1997, 54 per cent of the women in the US labour force would have had to change occupations to be distributed across occupational categories in the same proportions as men (Jacobs, 1999: 130). The particular forms of labour-market segmentation and their implications for household livelihoods and regional economies vary from place to place, but women's lower status in the hierarchy of occupations, their lower levels of remuneration and their reduced opportunities for advancement compared to men are a common theme. A substantial gender wage gap

161

remains, due in large part to job segregation; in 2007, for example, women working full time, year round in the US earned 80 per cent of men's earnings (English and Hegewisch, 2008).

Economic geographers have probed this theme along two main intersecting lines, namely from the perspective of employers and from the vantage point of individuals embedded in households and communities. Because of their association with domestic responsibilities, women have been viewed by potential employers, by their male partners and often by themselves as secondary wage earners, that is earners whose wages are supplementary to those of the male primary breadwinner. This secondary status, along with other cultural norms traditionally linked to femininity, such as manual dexterity, lack of physical strength, passivity and submissiveness, have marked women as cheap, docile labour, particularly suited for certain kinds of repetitive tasks and attractive to capital in certain situations.

Economic geographers have shown that this view of female labour, which treats women as an undifferentiated, homogeneous mass, masks the great variability of processes occurring in different geographic contexts. Within the industrialized world, Nelson (1986) demonstrated that employers located routine, back-office work in the middle-class suburbs of the San Francisco Bay area in order to attract the part-time labour of the well-educated, white women living there who did not want to travel far to work outside the home; England (1993) documents a similar process in Columbus, Ohio. Similarly, Hanson and Pratt (1992) found that in Worcester, Massachusetts, manufacturing and producer services employers were locating their facilities close to the residential areas housing particular kinds of female workers (e.g. immigrant Latinas for work in an industrial laundry in the central city; wives of blue-collar men for work in textile mills in towns outside of the city). Attention to gender in studies like these has highlighted the importance of much more finely scaled gender/ethnic/spatial division of labour than had previously been seen and has shown that local labour market areas are far smaller in geographic extent for some groups of people, especially some women, than had previously been realized. Gender and racial differences in the commute to work, a measure of spatial access to employment, are discussed in 2.3 Accessibility.

Similar processes operating at an international scale, but again playing out differently in different places, shed light on the globalization of capital flows (see 4.2 Globalization) and the development strategies of states in the global south (see 3.1 Industrial Location). The *maquiladoras* clustered along the US–Mexico border are a familiar example of international capital locating manufacturing facilities with the explicit intention of employing large numbers of women for low wages. Meier (1999) describes how the government of Colombia pursued cut-flower production as a development strategy, premised on the availability of cheap, largely female labour, drawn into the paid workforce to undertake the difficult and often dangerous work of tending vast fields of roses. Mullings (2004) cautions, however, that, absent effective institutional structures, economic activity based in large part on international capital and poorly paid women workers, as in tourism and travel in the Caribbean, is not an effective development strategy.

Feminist geographers have documented how gender relations within the workplace and within the household help to create and sustain these gender-based labour-market inequalities. McDowell (1997) and Jones (1998) have shown how the male-dominated and male-oriented daily social practices within workplaces in the financial-services sector in the City of London create a culture that is inimical to, and devalues, women. Gray and James (2007) examined practices within information-technology firms in Cambridge, UK, to learn how they differentially affect women's and men's ability to contribute to their firms' learning and innovation process and therefore to their firms' competitiveness. These studies, undertaken in fields with highly educated, professional workers, reveal myriad workplace practices – ranging from work hours and manner of dress to patterns of speech – that, however unintentional, limit women's opportunities at work.

163

Gender relations within households and communities also work to construct many women as secondary workers. To the extent that women retain primary responsibility for caring work in homes and communities, they are disproportionately likely to work in part-time jobs, which, as Gray and James (2007) point out, make full engagement in workplace interactions difficult. In an African agricultural context, Carney (1993) and Schroeder (1999) have documented how gendered social relations within household and community in Gambia have shaped agricultural practices there; in Carney's case, the norm that women should tend to subsistence crops and men to cash crops meant that men took control of rice production, which had been led by women, when the introduction of irrigation led to its becoming a commodity crop.

Most of these studies have demonstrated how the boundaries between productive and reproductive activity are indeed blurred, with each realm of work affecting the other in countless ways.[4] This exposure of the interdependencies between production and reproduction (or between home and work) in people's everyday working lives – an exposure that is the result of attention to gender – suggests that much of economic geography concerns livelihoods (see Sheppard, 2006). Research in many geographic settings by feminist economic geographers has detailed the variety of ways that people's livelihood strategies constantly engage the interface between production and reproduction. Among such studies are Pavlovskaya's (2004) investigation of the Moscow economy during the transition from socialism to capitalism, Hapke and Ayyankeril's (2004) examination of fishers and fish traders in South India, Wong's (2006) detailed look at the remittances that Ghanaian immigrants in Toronto, Canada, send to family members in Ghana, and Mattingly's (2001) study of Latina domestic workers in San Diego, California. These studies provide strong evidence that understandings of productive activity are impoverished and misleading unless they are rooted in and accompanied by understandings of linkages between productive and reproductive realms.

The centrality of these linkages to economic geographic processes has led many feminist economic geographers to focus on the networks of social relations that knit together diverse dimensions of everyday life (home, workplace, community, state) (see 5.5 Networks). Because social relations are deeply gendered, networks function differently for women and men (as well as, of course, for different groups of women and different groups of men). In their study of urban women workers in Indonesia, for example, Silvey and Elmhirst (2003) showed the impact of gendered expectations on these young women's lives; through the networks joining these women, as daughters, to their families in rural areas, family members placed demands on their wages that benefitted the rural households to the detriment of the urban women wage earners. Similarly, Wong (2006) describes the power of transnational networks to transmit gendered expectations from Ghana to Canada. Another way in which networks are gendered is that women find themselves excluded from the more powerful networks and from the most powerful people within a network (McDowell, 1997; Silvey and Elmhirst, 2003; Gray and James, 2007)

Each of these investigations considers how gendered power works through context-specific discourses and practices. By documenting how gender is lived differently in different times and places, these studies together also highlight that gender structures are fully capable of being transformed. One transformation that is currently receiving attention

164

is that of masculinity and masculine identities. McDowell (2005), for example, explores how the restructuring from manufacturing to service economies in the UK is affecting traditional views of masculinity by requiring men in service jobs to adopt characteristics that have traditionally been associated with femininity (e.g. deference, attention to appearance).

KEY POINTS

- Gender refers to perceived differences between women and men and to the unequal power relations that flow from these perceived differences. Because the meanings and practices of gender emerge through everyday interactions, they vary from place to place; geographic context is therefore central to understandings of gender, just as gender is central to understanding economic geographic processes.
- The ways in which scholars have thought about gender have shifted over the past three decades from seeing gender in terms of a universal male–female binary to seeing gender as a social construction; the latter view highlights the importance of geography.
- Gender has entered economic geography primarily through studies of work. These studies, many of which have documented the importance of investigating the linkages between productive and reproductive activities in different geographic contexts, have altered explanations of economic restructuring and geographic change.

165

FURTHER READING

Lawson (2007) provides an overview of caring work. Schroeder (1999) analyzes how gendered household and community relations in Gambia affected, and were affected by, changing agricultural practices

NOTES

1 Not all women in the field, however, explicitly focus on gender in their work.
2 Feminist economic geographers share with feminist economists (e.g. Nelson, 1993) an interest in households' and communities' everyday activities that influence livelihoods.

3 As Fairclough (1992: 64) notes, 'discourse is a practice not just of representing the world, but of signifying the world, constituting and constructing the world in meaning'. In other words, the language, texts, material objects and representations used to communicate information and guide social action also help to construct social identities, social relationships and widely recognized beliefs and forms of knowledge. In the case of gender, these social constructions – manifest in such things as occupational norms, product advertisements and expectations about workplace behaviour – can play a central role in perpetuating inequalities and stereotypes.

4 Although close linkages between productive and reproductive realms have perhaps been accepted as the norm in subsistence agricultural communities (see Seppala (1998) and Bryceson (2002) for discussions of subsistence livelihood strategies in Arica), they were not acknowledged by scholars of industrialized societies, whose whole premise was based on the spatial separation of home and work.

5.3 INSTITUTIONS

Institutions are patterns of behaviour that structure societies and make everyday life more consistent and predictable. They are manifest as organizations (e.g. the World Bank), laws and regulations, socio-cultural traditions (e.g. marriage), and in the formal and informal rules, norms and conventions governing socio-economic activities. Institutions coordinate economic activities by encouraging or discouraging particular patterns of behaviour in individuals, firms, and state actors and research on institutions provides an important perspective on the development of regional and national economies over time. For economic geographers, institutional analyses focus on how institutions help or prevent regions from growing and innovating, how institutions create and sustain uneven geographies of socio-economic opportunity, and about the institutional challenges concomitant with global economic integration.

Institutions and the historic development of capitalism

Institutional theory emerged in the early twentieth century when scholars such as Max Weber and Thorstein Veblen began to critically examine the evolution and power of modern capitalism's organizational structures (e.g. firms, markets, property rights). Weber (1905, 1958) documented the historic rise of western forms of capitalism arguing that transformations to economic, political and religious institutions drove the process of economic modernization. Economic innovations in accountancy, currency, production technologies, incorporation and property rights enabled the transition from rent-oriented to profit-oriented economies. Political changes, namely the shift from the liturgical state (i.e. a feudal system with servile citizens or subjects) to the tax state (a system with free subjects who serve as sources of tax revenue) governed by rational law and an industrial bourgeoisie, helped to separate economic power from political power. These political changes were aided by transformations in religious institutions (e.g. the Protestant work ethic, the concept of vocation) that supported the social-class structure and which encouraged communities to be more open to trade with outsiders (Weber, 1905). Taken together, these changes created institutional landscapes that rewarded individuals by following their rational-economic self-interests.

For Veblen (1925), the institutions of capitalism vary culturally and geographically making it impossible to achieve a universal set of rules and norms to govern the world economy. He argued that institutions were mental habits that evolve or 'drift' in a society over time through the influence of a complex set of intervening contextual factors such as history, cultural traditions and societal values. In his analysis of Japan, for example, Veblen (1915) argued that the conflicts between traditionalism and the logic of the business enterprise system (modernity) had created a uniquely 'quasi-feudalistic' set of economic institutions well positioned to borrow and integrate modern industrial technology. Veblen further argued that this unique institutional blend made Japan a very powerful nation state.

168 Institutions and regional development

Institutional thinking in economic geography emerged in the 1990s in response to the work of a new generation of institutional economists (e.g. Williamson, 1985; North, 1990), evolutionary economists (e.g. Nelson and Winter, 1982), and economic sociologists (Powell and Dimaggio, 1991). Moving beyond Veblen's view of institutions as mental habits, these scholars conceptualize institutions as structures that organize economies by providing rules and guidelines that govern how actors behave in business relationships. These patterns are constituted and 'carried' by agents (e.g. individuals, firms, workers and state actors) whose day-to-day identities, perceptions, decisions, and power struggles are embedded in them (Jessop, 2001; Scott, 1995) (see 5.4 Embeddedness). When effective, some institutions can work to prevent market failures (e.g. pollution), facilitate innovation, reward risk-taking and foster entrepreneurship whereas others can obstruct development, prompting societies to continuously modify their institutions. Institutional research in economic geography originally emphasized formal institutions (e.g. industrial promotion systems, state agencies) and generally recognized that institutions such as growth coalitions, and organized labour play a key role in fostering innovation and directing industrial restructuring processes (Amin, 1999). More recently scholars have focused on informal institutions, such as those associated with particular cultures, and their role in industrial and regional development (see 5.1 Culture).

Institutions are critical for regional development, and economic geographers have studied institutional change, and in particular, how countries adjust their institutions to meet the demands and consequences of economic globalization (see 4.2 Globalization). Florida and Kenney's (1994) study of Japan's post-war industrial restructuring took the case of automobile-production systems and showed how institutions combined with organizational innovation enabled Japanese firms to outcompete European and American enterprises. In this case, institutions evolved through struggles among managers, workers and the state that led to new forms of industrial organization based on flexibility, just-in-time production and technological innovation.

For Amin and Thrift (1993), regions need to create institutional thickness in order to maximize their innovativeness and growth potential. Thickness is achieved when a region's institutions: a) effectively serve diverse roles (e.g. worker-training organizations, investment-promotion agencies); b) are collectively legitimized and recognized by politicians, businesspeople and workers; c) discourage behaviours (e.g. corruption, cheating) that might impede growth; and d) help create dense networks of social relations among economic actors. It is through this diversity, efficacy and these relationships that institutionally 'thick' regions embed specialized knowledge locally, knowledge that provides long-run competitive advantages vis-à-vis industries located in other places. In applying this concept to the study of the UK's Motor Sports Valley, Henry and Pinch (2001) found that informal institutions, such as those that encourage labour turnover and information sharing among small enterprises in the cluster, which in turn help create specialized technical expertise and play a key role in ensuring the region sustains its position as a global centre for innovation in the motor sports industry (see 2.1 Innovation).

Studies such as these demonstrate why it is important for regions to develop relational assets that can support the evolution of institutions that effectively and flexibly manage innovation. Relational assets are embodied in the networks that link economic actors, are enabled by trust, cooperation and reciprocity, and can lead to the creation of untraded interdependencies in a place (Storper, 1995, 1997). This

169

'human infrastructure', when coupled with appropriate communications, logistics and manufacturing infrastructure, effective capital allocation and innovation-driven forms of industrial governance, can transform a regional economy into a learning region with 'knowledge-intensive' forms of industrial organization (Florida, 1995: 534). However, the learning region concept remains contested and is difficult to operationalize in research. For example, the institutions are considered 'right' when these regions are able to access, absorb, create and diffuse new knowledge and build learning capacity among their firms and workers (Morgan, 1997).

Institutions and spatial inequality

Institutions create unequal distributions of economic opportunity within and between regions. Regional institutions discourage investments and innovation by increasing the costs and perceived risks of entrepreneurial behaviour (Yeung, 2000). For example, Putnam's (1993) study of Italy's regional development demonstrates how inflexible, corrupt and excessively hierarchical institutions in the Calabria region prevented development while the northern region of Emiglia-Romagna successively industrialized through significant support from regional institutions that promoted trust, reciprocity and innovative risk taking. In other words, northern Italian institutions promoted and rewarded innovation and industrial development through their contributions to the region's social capital (see 5.5 Networks).

Within regions, institutions can also create and sustain socio-economic inequality. As Hudson (2004) observes, economies are socially constructed and 'instituted' by rules and understandings about 'proper' behaviour and conduct. When individuals challenge or fail to comply with existing rules, they may cease to be 'legitimate citizens' and institutions often discriminate against members of society based on gender, class, race and/or ethnicity. This discrimination, in turn, may lead to highly uneven landscapes of economic opportunity. For example, Gray and James's (2007) study of Cambridgeshire's high-tech cluster demonstrates how institutions in a 'successful' regional economy can produce and reproduce gender inequality while Blake and Hanson (2005) show how financial institutions in the USA discriminate against female

170

entrepreneurs and thus limit their abilities to develop their enterprises (see 5.2 Gender).

Institutional diversity and the challenge of global economic integration

The diversity of institutions in the world economy creates barriers for interregional trade, investment and knowledge flows. International financial institutions (IFIs) such as the World Bank and International Monetary Fund view such diversity as a threat to economic globalization and encourage countries to commit to a common set of market institutions aimed at reducing the complexities and costs of transnational business. These policies focus on creating market-friendly or neo-liberal institutional landscapes that protect private property rights (e.g. for land, investments and intellectual property), enforce formal business contracts, encourage market-oriented solutions to social and environmental problems, improve information flows, reduce corruption, and reward free enterprise and entrepreneurship (World Bank, 2001) (see 1.3 State). Despite their proliferation, however, these global institutional measures have only partially integrated the world economy, and many scholars believe that they disproportionately favour core regions and centres of power (e.g. Washington, DC) at the expense of peripheral places, non-western peoples and poor communities (Peet, 2007) (see 4.1 Core–Periphery).

171

Institutional differences between places can also play a significant role in knowledge exchange and transfer. For Munir (2002), the success of international technology transfer programmes hinges on the degree to which institutional frameworks in receiving countries are compatible with technology oriented institutions in transferring countries. In other words, institutional proximity or convergence plays a key role in determining the extent and quality of transnational knowledge flows (Gertler, 2001). Gertler et al.'s (2000) study of Canada's high-technology industries demonstrates how the institutional barriers discourage foreign firms in Canada from building learning and knowledge transfer relationships to local Canadian firms. This relative isolation is significant in that it reduces the knowledge diffusion or spill-over effects that stem from inward FDI in Canadian provinces.

New developments and the wider relevance of institutional thinking

Two recent theoretical perspectives in economic geography engage with institutions and their role in regional and industrial development processes. The relational approach focuses on the role of agency and power in driving institutional change and determining the influence institutions have on economic action. In this framework, institutions not only condition economic activities, but also play a key role in shaping the identities of individuals such as workers, entrepreneurs and business executives (Jessop, 2001). Moreover, institutions are recognized as dynamic structures that evolve over time and in space through interactions between economic actors, their material practices, geographical contexts and power structures (Bathelt, 2003, 2006). Contextual factors such as a region's culture, history and dominant political ideology play an especially important role determining the direction and characteristics of institutional change. By understanding how these actors, practices and factors legitimize and stabilize context-specific institutions, relational economic geographers hope to develop more dynamic theories of regional development and economic globalization (Yeung, 2005).

172

A second, and related, new direction takes an evolutionary perspective on the development of regional economic institutions and economic development trajectories (Boschma and Frenken, 2006). Evolutionary approaches focus on the role of institutions in guiding economic action, driving technological change and structuring interactions among labour, entrepreneurs, industry leaders and the state (Nelson and Winter, 1982; Essletzbichler and Rigby, 2007; Truffer, 2008). Institutions shape these routines, know-hows and regimes, and play a key role in guiding industrial and technological evolution.

Importantly, institutions are widely recognized to play an important role in shaping other aspects of a region, place, or industry's economic geography. Institutions structure gender, intra-firm and labour relations, organize global value chains and networks, serve as social contexts wherein economic activities are embedded and act as a central mechanism through which states govern economic activities. As such, the relevance and importance of the concept is evident in other chapters of this book.

KEY POINTS

- Institutions are stabilized patterns of behaviour that structure societies and make everyday life more consistent and predictable. They are manifest as organizations (e.g. the World Bank), laws and regulations, socio-cultural traditions (e.g. marriage), and in the formal and informal rules, norms and conventions governing socio-economic activities.
- Institutions play a key role in regional development processes by organizing production activities and facilitating or obstructing learning and innovation.
- Institutions create spatial inequality by enabling or empowering particular regions, communities, social groups and/or forms of knowledge.
- Inter-regional differences in institutions act as barriers to global flows of knowledge, capital and labour within and between firms and markets.

FURTHER READING 173

Rodríguez-Pose and Storper (2006) discuss the institutional complementarities between communities and the societies that contribute to regional development processes. Sunley (2008) critiques relational economic geography and argues for more historical and evolutionary forms of institutional analysis. A recent special issue of *Economic Geography* (Volume 85, Issue 2, 2009), focusing on evolutionary economic geography debates, analyzes the role of institutions in the evolution of industries and economies.

5.4 EMBEDDEDNESS

Embeddedness refers to the notion that economic activities are inseparable from social, cultural and political systems. Although the concept was first developed by sociologists, economic geographers have actively used it to study how spatially and historically situated non-economic factors influence the development of firms, industries and regions. Primary focus has been placed on understanding how embeddedness influences innovation possibilities, shapes market relationships, reproduces inequalities and creates path dependencies in regional development. Recent debates have raised concerns as to whether the concept excessively privileges the local, and whether it can be adequately operationalized in empirical research.

The foundations of embeddedness

Embeddedness is generally credited to the work of two scholars – Karl Polanyi and Mark Granovetter – who viewed the concept from different perspectives. Polanyi's (1944) landmark study *The Great Transformation* was the first to articulate and popularize the concept of embeddedness. Polanyi argued that mid-nineteenth century Great Britain underwent a radical reorganization when economic liberals sought to transform the country from a protectionist state-centred economy into a free-market society based on the principles of laissez-faire. The goal was to disembed the market from cultural, social and political systems and then to restructure these systems into a self-regulating market and a market society (Krippner and Alvarez, 2007). Polanyi believed that this transformation would inevitably lead to social problems (e.g. labour exploitation, poverty, inequality) which would, in turn, force the state to interfere with the free market system. Polanyi's influence is visible in contemporary studies that examine how states adjust social policies in response to economic crises (e.g. Kus, 2006), that analyze how states strive to remain autonomous from, yet able to guide, economic activities (e.g. Evans, 1995), and in the varieties-of-capitalism literature which, in part, demonstrates how political-economic systems can be differentiated with respect to the degree of market freedom (e.g. Crouch and Streeck, 1997) (see 1.3 State).

More influential for most economic geographers is the work of Mark Granovetter, who focused on the networks of personal relationships linking people, firms and places, and argued that all economic action is embedded in networks. Although these networks are shaped by larger-order institutions, their purpose, structure and value is determined primarily by individual social interactions. Because all economic activities are embedded in networks of some kind, the evolution of firms, industries and economies is best understood through network analyses (see 5.5 Networks). In arguing for a focus on networks, Granovetter (1985) sought a middle ground between what he considered 'oversocialised' accounts of economic action put forth by sociologists (e.g. Parsons and Smelser, 1956) and 'undersocialised' rational-choice accounts preferred in institutional economics (e.g. Williamson, 1985).

Granovetter's critique was extended through sociological studies that analyze the influence of intra- and inter-firm networks on performance and innovation in organizations and industries. Embedded or bonding network ties are essential for creating reliable and stable business relationships, but entrepreneurs and firms also need weak ties that facilitate access to new sources of information and/or individuals (Granovetter, 1973; Burt, 1992; Uzzi, 1996).[1] Embeddedness also influences consumer choices (Dimaggio and Louch, 1998), the livelihood strategies of immigrants (Portes and Sensenbrenner, 1993), and the inter-state relationships governing global trade flows (Ingram et al., 2005).

175

Embeddedness and economic geography

Economic geographers' approach to the concept, although similar to that of sociologists, is distinguishable by its emphasis on place or region-specific forms of embeddedness. Embeddedness in a place can be measured in a number of ways – length of residence, the density of interpersonal or inter-firm networks, the level of trust in relationships, and/or based on the role of local cultural, social or political institutions in structuring or guiding business activities. An individual, firm or industry's 'rootedness' influences: a) the development of industrial districts and clusters; b) the evolution of regional economies; c) the socio-spatial dynamics of innovation; d) patterns of transnational trade and investment; and e) the spatial structure of labour markets.

Building on Granovetter's (1985) argument that industries constituted by small firms embedded in dense networks would be better able to compete against large firms, Harrison (1992) examined the role of embeddedness in the development of flexible production systems and industrial districts. He argued that embeddedness played a central role in the success of industrial districts (e.g. Silicon Valley, northern Italy) and that this embeddedness was achieved through shared experiences, trust and cooperative competition ('coopetition') through inter-firm networks. These networks and trusting relationships can reduce the transaction costs and create Marshallian externalities such as knowledge spill-overs and specialized labour pools (see 3.2 Industrial Clusters). Despite their innovative potential, Harrison cautioned against romanticizing the power in networks and industrial districts. This is because industrial district relationships may be sustained through racial, class and/or cultural homogeneity, and the 'systematic superexploitation' of women and immigrant labour. In addition, large corporations can benefit significantly from the intensive, cost-reducing competition among the small-scale suppliers located in industrial districts (Harrison, 1992, 1994).

176   Regional development is also shaped by the embeddedness of economic actors in localized social, cultural and political institutions. Grabher (1993) demonstrated how embeddedness caused different types of industrial 'lock-in' in Germany's Ruhr region. Functional lock-in occurs when firms are embedded in highly structured and interdependent sets of supply-chain and market relations that are difficult to change. Cognitive lock-in occurs when the thinking of industrial and regional leaders becomes overly embedded in perspectives that prevent new ideas from gaining legitimacy or power. Political lock-in occurs if the coalitions and institutions governing and supporting industries excessively favour traditional or 'sunset' industries at the expense of new sectors or 'sunrise' industries (see 1.3 State). Lock-in situations have been identified in other places (e.g. Eich-Born and Hassink, 2005; Hassink, 2007) and these create particularly significant challenges for policy makers striving to develop new industrial sectors.

Embeddedness also influences how firms produce, use and diffuse tacit forms of knowledge. Tacit knowledge is embedded in geographic contexts, cultures and/or industrial communities; is articulated or realized through actions and experiences; is best transferred through spatially proximate relationships; and is an important means through which clusters and regions develop localization economies and global competitive advantages (Polanyi, 1967; Gertler, 2003; Morgan, 2005) (see 2.1 Innovation; 3.2 Industrial Clusters). A particularly significant strand of this research focuses on the tacit knowledge that is embedded in communities of practice (CoP), 'constellations of diverse learning communities' within or between firms such as those associated with particular industrial sectors (e.g. biotechnology) or administrative activities (e.g. insurance-claim handling) (Wenger, 1998; Amin and Cohendet, 2004; Amin and Roberts, 2008). Participants in these communities create and distribute knowledge through both the formal (e.g. contract driven) and informal (e.g. casual or social) relationships that structure their day-to-day activities.

A fourth direction in embeddedness research examines its influence on international trade and foreign direct investment (FDI). Bellandi (2001) observed that when firms interact with foreign places they can either remain unembedded in the local economy, and thus transfer little knowledge and resources from their home economies into the local economy, or they can develop an 'embedded unit' able to sustain intensive socioeconomic interactions locally that may lead to increased knowledge and resource flows into the locale. Achieving embeddedness at a distance poses challenges both for investing firms and investment-receiving regions as each must balance its internal or local priorities with a recognition of, and support for, the others' needs. Moreover, power matters in that it can determine who controls the embedding process. For example, the Vancouver film industry has historically been dependent on FDI from the USA and is thus partially embedded in US production standards and practices (Coe, 2000). In contrast, the Chinese government has 'obligated' foreign investors in the automobile industry to embed their activities in local supply chain networks if they want access to the lucrative Chinese market (Liu and Dicken, 2006).

177

Finally, economic geographers use embeddedness to assess the role that an individual's socio-spatial rootedness plays in her/his employment opportunities, entrepreneurial strategies and/or livelihood activities. Individuals become embedded in places through their residence over time, the extent and quality (e.g. trustworthiness) of their personal contacts, and by the manner in which social relations based on family, culture, class and/or gender structure the kinds of occupational or business networks available to them (Hanson and Pratt, 1991; Hanson and Blake, 2009). Such forms of embeddedness play an important role in determining how a particular social group's (e.g. women, immigrants) access to opportunities is constrained by structural inequalities (see 2.3 Accessibility).

Critiques of embeddedness

Although embeddedness is widely accepted as an important concept in economic geography, there have been a number of important methodological and theoretical critiques in recent years. These concerns and debates have focused on how to better operationalize the concept in empirical research, about the scale at which studies of embeddedness are typically carried out and, most broadly, about whether the concept remains useful for theories in economic geography. Methodologically, Mackinnon et al. (2002) criticize an absence of clear empirical connection between embeddedness and economic outcomes (e.g. learning, innovation, agglomeration). Because individuals are embedded in multiple and overlapping sets of social relations, collecting data on embeddedness through interviews is difficult to accomplish (Oinas, 1999; Ettlinger, 2003). As such, a significant challenge for researchers is to develop measures and indicators that precisely determine the role of embeddedness in economic activities.

178

A second critique relates to the scale of embeddedness research. Hess (2004) contends that studies of embeddedness are typically too focused on local activities and thus provide only a partial understanding of its role and constitution. While these territorial forms of embeddedness – an actor's (e.g. firm, entrepreneur) connection to a particular place or region – are one significant dimension, it is also critical for geographers to understand how societal and network forms of embeddedness influence economic activities. Societal embeddedness relates to the larger-scale cultural, political and social institutions and attributes that shape actors' identities and understandings of the world. Network embeddedness describes the social relationships that actors are shaped by and embedded in (see 5.5 Networks). Hess argues that all economic actors (i.e. firms, workers, consumers) are uniquely embedded in institutions or structures at multiple scales, and that this heterogeneity plays a significant role in regional development processes and global trade and investment relations. For example, transnational retail firms' concerns about their image at home (i.e. their societal embeddedness) may discourage them from investing in territories or places where child labour is the norm (Hughes et al., 2008). Alternatively, as Izushi (1997) demonstrates in the case of Japan's ceramics industry, if the activities of a local firm are principally embedded in relationships to key external buyers, this may prevent the firm from creating knowledge spill-overs or externalities for the local economy.

Finally, there are those who broadly question whether the concept has outlived its utility. Peck (2005) argues that studies of embeddedness over-emphasize the immediate social context at the expense of structural factors like class relations. For Jones (2008), embeddedness provides at best a limited understanding of the spatial processes driving development and it conflates economic processes (e.g. networking) with economic outcomes (e.g. growth). He further argues that economic geographers would be better served by analyzing socio-spatial practices that drive economic outcomes, not just the contextual realities. Despite these critiques, embeddedness remains an important conceptual lens through which geographers examine economic actions situated in place, region and space.

179

KEY POINTS

- Embeddedness refers to the notion that economic activities are inseparable from social, cultural and political systems. Although the concept was first developed by sociologists, economic geographers have used it extensively to study how spatially and historically situated non-economic factors influence the development of firms, industries and regions.
- Individuals, firms and industries become embedded in places and networks through residency or spatial fixity, inter-personal or inter-firm relationships, mutual trust, shared experiences, cultural homogeneity, unequal power relations, structural interdependencies and inequalities, and/or through institutional forces that constrain economic opportunities and possibilities.
- Studies of embeddedness have focused on its role in shaping industrial districts and clusters; its positive and negative contributions to regional economic development, learning and innovation; its influence on transnational trade and investment relationships; and on how it structures the economic opportunities available to particular social groups (e.g. women, immigrants).

FURTHER READING

Gertler et al. (2000) examine the challenges of successfully embedding FDI in a regional economy. Glückler (2005) evaluates how embedded social practices shape international consulting markets. James (2007) details the cultural embeddedness of a high-tech industry in the USA. For a recent analysis of regional lock-in through embeddedness, see Lowe (2009) on Mexico's apparel industry. For a recent discussion on the limitations on the embeddedness concept, see Hervas-Oliver and Albors-Garrigos (2009) analysis of how innovation occurs in an industrial cluster.

NOTE

1   Embedded ties are strong relationships between individuals built through trust, socio-economic interdependencies, cultural affinity, power relationships and/or shared experiences. Weak ties generally have few of these characteristics but can provide new and novel connections (i.e. bridges) to different kinds of people, organizations and places.

5.5 NETWORKS

Networks are socio-economic structures that connect people, firms and places to one another and that enable knowledge, capital and commodities to flow within and between regions. The concept helps explain how economic activities are organized across space and how economic relationships (e.g. between firms, businesspeople) influence growth and development in places. Economic geographers have principally studied networks from two perspectives. The first emphasizes how networks organize industrial clusters and global trade and investment relationships. The second approach focuses on how individuals and firms construct and use social networks in the economic sphere, and on how access to networks is shaped by social inequalities related to gender, class and/or ethnicity.

Networks as economic organization

Networks differ from markets and hierarchies in that they help to organize economies through socio-economic relationships, not price-setting mechanisms (markets) or the imposition of power (hierarchy) (Powell, 1990; Powell and Smith-Doerr, 1994). While price and power may play a role in determining who participates, and how, in a network, the emphasis of network research is on how inter-firm and inter-regional relationships develop and how the structural characteristics of network ties (i.e. their strength or inequality) shape regional development and innovation processes. Geographers have primarily studied the form, function and influence of networks at two scales – the regional and the global.

At the regional scale, network research focuses on how networks contribute to agglomeration economies and the formation of clusters and industrial districts (see 3.2 Industrial Clusters). Learning is an important characteristic of successful clusters, in which inter-firm networks play a central role in the creation and diffusion of new knowledge (Camagni, 1991). Networks contribute to knowledge creation and information diffusion through two mechanisms, one through networks creating 'buzz' or flows of information within the cluster through spatially proximate relationships between employees, firms and state agencies (Bathelt et al., 2004). Second, networks establish 'pipelines' between local and non-local firms to exchange information and knowledge. Amin and Thrift's (1992) analysis of 'neo-Marshallian nodes' such as industrial districts in San Croce (Italy) and London (UK) showed how the success of these regions depended significantly on their ability to link into global corporate networks. In order to build ties to these districts, regional policy makers needed to create institutional conditions that improve knowledge, resource and capital flows (Amin, 1999; Coe et al., 2004) (see 5.3 Institutions). For developing countries, global–local linkages such as these can significantly affect the prospects for development, and network analyses can explain how and why some places and people benefit disproportionately from transnational trade and investment relationships.[1] For example, Bek et al. (2007) demonstrate how various industry and development actors help shape the network linking UK wine consumers to the labour practices of South African wine producers.

For Castells (1996), the global economy is a 'network of networks' where information creation, access and application play a central role in determining which places and regions are most productive and competitive. In the age of information-driven capitalism, a firm or region's competitiveness is based in large part on its ability to connect to global networks where capital and information (e.g. about prices, innovations and markets) flow. This global 'space of flows' is an outcome of new information technologies (IT), post Cold War political reforms (e.g. multi-lateral trade agreements) and the New International Division of Labour (NIDL) that developed through the offshoring, outsourcing and OEM practices of MNCs (see 4.2 Globalization; 6.1 Knowledge Economy; 4.4 Global Value Chains).

Importantly, the NIDL is constituted through a global network of cities within which corporate functions, international finance firms, producer services, and communication and transportation systems are concentrated and/or controlled. As Friedmann (1986) observed, these cities play a key role in driving uneven global development and they are organized in a spatial hierarchy based on their functional role and power in coordinating global economic activities (see 4.1 Core–Periphery). The most powerful cities – primary world cities (e.g. London, New York, Tokyo) – serve as centres for capital accumulation as well as for the global trade and investment activities of MNCs. Primary world cities are networked with secondary and tertiary world cities (e.g. Sao Paulo, Hong Kong, or Houston) that serve as centres for particular industries and/or which provide advanced producer (e.g. legal, accounting, insurance, logistics) and financial services to MNCs operating in their regions (Knox and Taylor, 1995; Beaverstock et al., 1999; Taylor and Aranya, 2008). For example, Hong Kong plays a central role in organizing capital flows into and out of China, and Houston is a global centre for the petroleum industry (e.g. see Rossi and Taylor, 2006).

Networks as social organization

The second strand of research on networks focuses on the social characteristics of individuals and firms. Work in this area has closely paralleled on-going work in economic sociology and organizational theory (e.g. see Granovetter, 1985; Powell, 1990; Powell and Smith-Doerr, 2005). Economic geographers have extended these ideas to understand how social networks influence small enterprise and regional development processes, how network structures reflect social inequalities and how networks evolve through social interactions between individuals.

183

Interpersonal and inter-firm networks play a key role in regional development through their ability to mobilize social capital – what Woolcock (1998: 153) defines as 'information, trust, and norms of reciprocity inhering in one's social networks'. When effective at creating and mobilizing social capital, networks can improve information and resource exchange within and between firms and foster entrepreneurship (Anderson and Jack, 2002; Nijkamp, 2003; Yeung, 2005) (see 2.2 Entrepreneurship). Such networks best contribute to wider-scale development if urban or regional institutions foster generalized forms of trust such that network gatekeepers (e.g. large firms, 'community entrepreneurs') are best able to interact with and diffuse knowledge to a wide variety of firms and policy makers (Malecki and Tootle, 1997). For example, Murphy (2002) and Mackinnon et al. (2004) demonstrate how informal networks of small and medium-scale enterprises in Tanzania and Scotland shape innovation, information diffusion and market access. Ideally, networked regions or associational economies emerge over time to facilitate innovation and learning (Cooke and Morgan, 1998; Malecki, 2000).

Social capital also has a dark side as network architectures reflect institutional, cultural, racial and social biases that exist in all societies

(Hanson, 2000). These biases can lead to the exclusion of qualified, competitive and/or innovative individuals and firms from industrial or labour networks while privileging those who already possess significant social capital as a result of their gender, race, ethnicity or social class. For example, Harrison (1992) argues that the inter-firm networks structuring successful industrial districts were based on cultural homogeneity, Turner (2007) demonstrates how cultural factors shape livelihood networks in Indonesia, and Hanson and Blake (2009) show how entrepreneurial networks are segmented along gender lines that can limit women's access to information, finance and other forms of support.

Actor-Network Theory

Actor-Network Theory (ANT) has had a significant influence on economic geographers' theorizations of networks. ANT emerged out of science and technology studies (especially the work of anthropologists and sociologists) and its early proponents were interested in better understanding the social, technical and material processes driving scientific discovery and research. ANT was principally developed by Bruno Latour, Michel Callon and John Law but was inspired and influenced significantly by Thomas Kuhn's (1962) study of scientific revolutions, Michel Serres' philosophy of science and a variety of philosophical traditions including phenomenology, post-structuralism, social psychology and ethnomethodology (Callon, 1986; Latour, 1991; Law, 1992, 2008; Knorr Cetina and Bruegger, 2002).

ANT scholars argue that it is impossible to separate economic actors from their actions, as the latter are embedded in multiform and multiscalar relationships. Under ANT, there is no such thing as an isolated actor (i.e. agency) or network (i.e. structure); instead, there are only actor-networks constituted by individuals, their diverse past and present relationships to others and material artifacts (e.g. technologies, texts, money, physical space) that are relevant for a particular economic activity. When actors engage in interactions (such as exchange, investment, research or knowledge communication), they mobilize their cognitive understandings by building trust in relationships, and by referencing these material artifacts (Murdoch, 1995; Murphy, 2006).[2] Power is articulated through these interactions, with the goal of effectively *enrolling* others into her or his actor-network. *Translation* is a means through which ideas are transmitted in the process of enrolment. Enrolment leads to an *alignment* of the perspectives of the actors and this, in turn, facilitates economic action (e.g. trade, learning) both locally and translocally. These micro-social interactions and negotiations construct economic spaces, interconnections and interdependencies (Murdoch, 1998).

As Callon (1986) first demonstrated in his analysis of the relationships between scientists, fishermen and sea scallops in a French fishery, the study of enrolment, alignment and translation processes can provide important insights into the social dynamics and power relations driving economic and ecological outcomes. For some economic geographers, studies such as this were attractive in part because of their attention to context-specific details and their sensitivity to the role of individual agency in shaping economic, industrial and technological-development processes. More recently, Callon (1999) and Callon and Muniesa (2005) argued that markets are socially and spatially ordered through the cognitive frames used by actors in calculating costs, assessing prices and carrying out exchange relationships (Berndt and Boeckler, 2009). For others, ANT has contributed significantly to debates on how network relationships are formed, maintained, and extended in space and time (Grabher, 2006). Relational proximity is a common theme in these studies, in that the concept of spatial proximity alone is insufficient to understand the contemporary geography of economic activities (Bathelt and Glückler, 2003). By creating social spaces amenable to translation and enrolment, actors can build stable ties within and across multiple spatial scales and effectively 'act at a distance' on one another (Dicken et al., 2001; Hess, 2004). For Sheppard (2002), ANT offers a useful framework for understanding how global networks are constructed and how they *position* individuals, firms and places in unequal power relations. Ideas from ANT have been applied to a variety of studies in economic geography including Smith's (2003) analysis of relationships between Slovakian clothing producers and European apparel markets, Truffer's (2008) conceptual discussion on the links between technologies, societies and regional development, and Kim's (2006) study of the networks linking foreign MNCs to South Korea's seed industry.

185

Critiques of networks

Although the network concept has resonated widely within economic geography, important critiques exist. First, some argue that networks are portrayed too positively and that network studies fail to address the structural forms of power and inequality they may perpetuate or reproduce (Peck, 2005). Second, there are concerns that network studies, particular those applying ideas from ANT, are too descriptive, case-study oriented and micro-sociological. Because of this narrow focus, some question the ability of such approaches to rigorously explain how larger-order social and economic phenomena (e.g. institutions) emerge in a regional or national economy (Sunley, 2008). Third, some argue that too much emphasis has been placed on the trusting bonds and strong ties linking actors in the space economy. Instead, there is a need to look beyond these to weak ties or what Grabher (2006) characterizes as ephemeral 'public moments' where new and novel social interactions occur. Finally, some argue that network studies are too *ad hoc* methodologically and that the measures of performance in firms, clusters or industrial districts are inadequately linked to measures of network structure, extent and stability (Staber, 2001).

KEY POINTS

186

- Networks are socio-economic structures that interconnect people, firms and places to one another and which enable knowledge, capital and commodities to flow within and between regions. The concept helps explain how industrial activities are organized across space and how economic relationships (e.g. between firms, businesspeople) influence growth and development in places.
- Economic networks organize inter-firm relationships, global value chains, and inter-regional linkages and provide a valuable perspective on how and why uneven development persists in the global economy.
- Social networks can contribute to innovation, entrepreneurship and regional development or exclude individuals on the basis of race, gender, ethnicity or class. Relational proximity based on trust, cultural similarities and/or mutual understandings helps actors in different places build long-distance networks.

FURTHER READING

Kingsley and Malecki (2004) assess how networks contribute to entrepreneurship and success in small business. Glückler (2007) conceptualizes network evolution and provides a typology of how networks structures influence a region's prospects for innovation and development. Taylor et al. (2009) examine the structure of global networks linking North and South through a study of air-passenger flows between 1970 and 2005. Hadjimichalis and Hudson (2006) critique the network concept arguing that it insufficiently deals with power inequalities.

NOTES

1 Development geographers too are increasingly using network analyses to study how developing economies are influenced by the transnational networks associated with labour migration, foreign-aid relationships and social movements (Bebbington, 2003; Henry et al., 2004).

2 Although many reject more extreme versions of ANT, where inanimate objects and animals are said to have agency equal to that of humans, the notion that material artefacts shape social encounters and their outcomes is an important insight recognized by most.

Section 6
Emerging Themes in Economic Geography

What does the future hold for the economy and the discipline of economic geography? The final section focuses on emerging concepts that represent economic trends of increasing significance in the twenty-first century.

Although the idea of a Knowledge Economy was conceived in the 1960s, this concept has been steadily gaining in significance as the basis for newly emerging sectors of economic activities has shifted from natural to knowledge-based resources. Knowledge economy has been viewed as the future in advanced industrialized economies that are searching for alternative sources of job creation as manufacturing jobs are lost. Recognizing the importance of knowledge has also led to the reconceptualization of workers, from labourer to innovators. It should be noted, however, that many analysts view the geographical separation between knowledge-intensive and labour-intensive activities as problematic, as in reality the two are highly inter-dependent. Can an economy survive by becoming purely knowledge-driven?

As with the knowledge economy, the Financialization of economies did not emerge overnight, but has been in progress since World War II and will gain more importance with the financial crisis of 2008. Some view global finance capitalism as offering unprecedented opportunity, whereas others see it as see it as involving unsustainably high risks. This form of capitalism calls for a new form of governance structure, one that fully incorporates the speculative and volatile aspects of the financial world.

Consumption is a relatively new focus for economic geographers, who have long been preoccupied with production. Cross-disciplinary fertilization and learning from other disciplines are therefore prominent in research on consumption today. Yet, as we shall show in this chapter, consumption also has deep historical roots in economic geography. Consumption is increasingly what drives the global economy, and the simultaneous homogenization and diversity in consumer tastes have multiple repercussions, not only in production, but in conceptualizations of class. Consumers today are also a political force and are increasingly mobilized to shape, for example, the working conditions of people on the opposite side of the world.

Finally, the concept of Sustainable Development emerged in the 1980s in response to increasing concerns about the degradation of environmental resources. While there is agreement that sustainable development requires a balancing of social, environmental and economic priorities, significant disagreements remain on how to best achieve this balance. 'Weak' sustainability proponents believe that market forces can best guide sustainable forms of development while 'strong' proponents believe that state intervention is required to balance energy and material flows and that there are finite limits to economic growth. Economic geographers in particular have focused on the sustainability of industrial systems and urban or regional development processes, the efficacy of environmentally friendly product certification systems, and the impact of global climate change on regional development and livelihood strategies.

6.1 KNOWLEDGE ECONOMY

Today, knowledge is considered among the most important resources, and an essential driving force that contributes to economic growth (see 2.1 Innovation). Traditionally, economic geographers and economists considered capital and labour inputs as primary drivers of growth, yet the problems of resource-rich poor economies in the periphery (the post-independence Third World) made it abundantly clear that natural resources alone do not translate into growth (see 4.1 Core–Periphery). Various alternative views have emerged, with some focusing on techno-logical innovation and others emphasizing neo-colonialism, industrial organization, state institutions and regulations, national and regional culture as playing the major role in economic growth. Those who stress the notion of knowledge economy cater to a view that knowledge, rep-resented by skill levels and creativity of the labour force, is the key driver of innovation and economic growth. Economists have been attempting to model knowledge as endogenous inputs without notable success, until Romer's (1986) seminal work on the new growth theory. Romer assumed knowledge as an input in production with increasing returns in order to show how endogenous technological change leads to long-run economic growth.

Policy makers in the core economies are concerned with ongoing dein-dustrialization and the need for new economic sectors as sources of employment creation, and they view occupations that are closely asso-ciated with the knowledge economy as far more resistant to offshoring (e.g. educators, physicians). Combined, creating knowledge economy became both a focus of research as well as a policy objective for many economies.

Emergence of the knowledge workers

Research on the role of knowledge in economic growth emerged as part of the stage theory of economic development (see 3.3 Regional Disparity). Economic progress has been understood as a shift from agriculture to manufacturing, with various speculations on the next stage of development.

Machlup (1962), an economist, predicted the emergence of the knowledge economy in the US economy, measured by the dramatic increase in the share of knowledge production. Machlup observed demand to shift from physical- to knowledge-producing labour. Machlup thus emphasized the significant role of the state in the production of knowledge through investment in education and research at government laboratories and universities.

Drucker (1969), a well-known thinker in business management, advocated the emergence of knowledge society, in which the foundation of our economy has shifted from manual to knowledge workers, and the main social expenditures have shifted from goods to knowledge. Similarly, prominent sociologists (Touraine, 1969; Baudrillard, 1970; Bell, 1973) advocated the coming of the post-industrial society, in which societies have evolved beyond the manufacturing-dependent economy to the service-dependent consumer society. They observed an almost universal shift from the dominance of blue-collar manufacturing work to white-collar service work in North America and Western Europe. Because of this intellectual tradition, initial research on the emergence of knowledge workers was strongly associated with the rise of the service sector, although it is well recognized today that many manufacturing activities (e.g. electronics, bio-pharmaceuticals) are knowledge-intensive, and their spatial organizations reflect the availability of highly skilled knowledge workers (see Cooke, 2001, 2007).

192

From services to knowledge-intensive services

Research on the shift of employment from manufacturing to services sought to explain why such shift seems almost universally observed. According to Fuchs (1968), three reasons for the growth of service-sector employment include: 1) rising incomes will contribute to the growth of final demand in services; 2) the growth in goods-producing services will increase intermediate demand in services through the division of labour; and 3) because productivity increases in services are typically lower than industries, the sector demands more labour. Research on the service economy flourished in the 1980s on both sides of the Atlantic (see Stanback, 1980; Stanback et al., 1981; Gershuny and Miles, 1983; Daniels, 1985). However, this research was also met with significant criticisms. One is that the conceptualizations of services have been found to be highly problematic (see Singelmann, 1977; Cohen and Zysman, 1987; Sayer and Walker, 1993); another criticism is that, because services were viewed as 'auxiliary' to manufacturing, they cannot serve as the engines of economic growth. Kuznets (1971) argued that services tend to be more dependent upon the macro-economic growth of a region and, in particular, the demand for services generated by the manufacturing sector. Hirschman called these relationships inter-sectoral complementarities, which are critical for economic growth as they generate pecuniary external economies.[1] Cohen and Zysman (1987) emphasized the geographic dimensions of inter-sectoral linkages. According to them, because manufacturing and services are closely linked, growth strategies that exclusively focus on services are ineffective. Research has also found that a considerable diversity in the manner and speed is observed in the most advanced industrialized economies in the shift from manufacturing and service employment (Castells and Aoyama, 1994; Aoyama and Castells, 2002). In addition, the applicability of stage-theory of development to developing country context has been questioned (Pandit, 1990a, 1990b).

Furthermore, many services are labour intensive, non-exportable with limited market potential, and have also been associated with lack of innovation and low-productivity growth (Miles, 2000). Yet, recent literature suggests that certain service segments are exportable[2] (Wood, 2005), and exhibits high productivity as well as employment growth (Beyers, 2002). Increasingly, the knowledge and service functions of selling a product, such as branding, advertising and financial packages, are being viewed as critical competitive advantages for manufactured goods (Daniels and Bryson, 2002; Lundquist et al., 2008).

193

As a result, economic geographers interested in analyzing the knowledge dimension of the service economy have increasingly limited their scope of research to advanced producer services, or knowledge-intensive services (KIS) (Gillespie and Green, 1987; Marshall et al., 1987; Warf, 1989; Daniels and Moulaert, 1991; Beyers, 1993; Clark, 2002). These services typically represent specialized services that demand significant expertise, primarily serving firms rather than consumers as clienteles, and include such services as accounting, advertising, business consulting, engineering (architecture, software programming, bio-engineering), legal and financial services. More recently, research has focused on the globalizing aspects of advanced producer services (Leslie, 1995; Jones 2005).

The new economy

The economic growth in the 1990s associated with the spread of IT was coined the new (or digital) economy. Broadly, the new economy refers to the various structural changes that contributed to economic performance of the latter 1990s. According to Beyers (2002), the new economy is the result of 'the revolution in electronic computational and communications capabilities that have stemmed from the development of transistors, semiconductors, and the myriad of applications that pervade every aspect of our lives' (p. 2). Some view the impacts of IT to be equivalent to the second industrial revolution; technology was infrastructural in its effects, which were not limited to intra-sectoral growth, but also cross-sectoral (adopted in the financial sector, in manufacturing, in retail and distribution sectors, and in consumption). Thus, IT played a significant role in boosting overall productivity of an economy.

Castells (1989) claimed the industrial mode of development is replaced with the informational mode of development. In this era, information-processing activities predominate in production, consumption and state regulation. Active entrepreneurship exploiting new opportunities brought about the new technologies resulted in the dot-com boom of the late 1990s, which drove up stock-market prices particularly in the United States. Venture capital has been known as an important source of funds for business start-ups (Florida and Kenney, 1988), but they became particularly visible during the dot-com boom by profiting considerably from IPOs (initial public offerings) of firms they funded. The new economy also created an active entrepreneurial culture, as exemplified by that in the Silicon Valley in the United States (Saxenian, 1994) (see 5.1 Culture). A region's ability to attract and retain IT entrepreneurs is considered an important indicator for the future of regions in the knowledge economy (Saxenian, 1994; Malecki, 1997; McQuaid, 2002).

The combined effect of the promise of productivity increases and active entrepreneurship allegedly constituted a new economic logic under capitalism. It should be noted, however, that some scholars have questioned the validity of the new economy as a defendable concept. For example, Gordon (2000) claimed that the Internet did not live up to other revolutionary technologies of the past in terms of economic impacts.

194

The dominant class in the knowledge economy

Thrift's (1997) interests in knowledge economy emerged in the context of the 'cultural turn' (see 5.1 Culture). To Thrift, academics and management consultants combined to produce a particular discourse of knowledge, which, in turn, affects global business organizations. Thrift called the hegemony of managerialist discourse, 'soft capitalism', to emphasize the role played by certain discourse in capitalism, and warned against 'harmful abstraction' that arises out of certain corporate culture, such as constant auditing and measurement for success (Thrift, 1998). This new managerialist discourse is generated through the new dominant metaphors of global economic uncertainties as well as the permanent discourse of emergency (Thrift, 1997, 2000). As a consequence, the global 'cultural circuits of capital' emerge, through training in MBA programs and nurtured by the greater mobility of managers (Thrift, 1998; Hall, 2009).

Castells (2000) conceptualized the 'space of flows', which is a 'material form of support of dominant processes and functions in the information society' (p. 442). He claimed that the space of flows defines the spatial form and processes, and is made up of three layers – a circuit of electronic exchanges, nodes and hubs, and the spatial organization of the dominant, managerial and cosmopolitan elites. As a result, the space of flows is 'asymmetrically organized around the dominant interests specific to each social structure' (p. 445). The cosmopolitan elites, however, are no longer limited to those from the advanced industrialized economies. According to Saxenian (2006), 'new argonauts' from India, Taiwan and China, through their international business and social networks, are increasingly important as agents of knowledge circulation at the global scale.

195

Florida (2002) believes that we are entering a new phase of capitalist development with creativity being the primary source of economic growth. He argued that the prospects of cities today depend much on whether they manage to attract and retain what he coined as the 'creative class', which comprises a broad range of professional and technical occupations that engage in creative work, including artists, journalists, college professors, scientists and engineers. His research found strong associations between the locations of growth and the creative class, and suggested the role of amenities (climate, recreational and cultural opportunities) and tolerance (measured, in this case, by the share of coupled gay households in the population). Florida therefore argued that search for tolerance and diversity of population shapes the future economic prospects of cities in the United States.

Whereas Florida's thesis has been highly influential in shaping urban policies, it has been criticized as ambiguous, particularly with respect to the causal links between creativity and growth (Marcuse, 2003; Glaeser, 2005). Peck (2005) has pointed out that Florida's focus on the highly mobile urban elites in the prevailing market ideology comes at the expense of other pertinent issues, including social polarization between the creative haves and have-nots, and a greater impetus to commodify arts and cultural resources as a result of Florida's creativity thesis. Storper and Scott (2009) believe Florida's thesis has led to the privileging of individual choice in locational decision primarily based on amenity, and cautioned against disregarding a long history of urban growth, which is primarily based on people-following-jobs, not vice-versa.

Geography of the knowledge economy

196   Geography of the knowledge economy largely follows the geography of knowledge workers. Highly educated workers are geographically concentrated to form innovative milieux (Aydalot, 1985), or more recently, learning regions (Florida, 1995) or regional innovation systems (Cooke, 2001; Asheim and Coenen, 2005). Some have been developed with intensive state interventions (e.g. Hsinchu Science Park in Taiwan or Zhanjian Science Park in Shanghai, China), while others emerged without explicit state mandates (e.g. Silicon Valley, USA) (Castells and Hall, 1994; Saxenian, 2007). Global cities, such as New York, London and Tokyo are major concentrations of advanced producer services (Sassen, 1991) (see 2.1 Innovation).

The onset of information technology (IT) has allowed globalization and segmentation, with the latter referring to certain service-sector activities dividing work between front-office and back-office operations. Back-offices typically conduct routine activities such as data processing for airlines and credit-card companies, and remote customer-service functions (e.g. call centres). In the United States, back-offices first moved to the suburbs following highly educated housewives willing to work part-time near their homes at low wages ('pink-collar ghetto') (Nelson, 1986). Today, back-offices are increasingly located in countries that offer English-language proficiency as well as lower wages ('international back-offices'), although some are returning to exurbs and rural areas because of wage increases in offshored locations (e.g. Ireland) or issues with customer service quality (e.g. India) (see 4.2 Globalization).

The geography of the new economy is further complicated by the emergence of cyberspace. Today, virtual spaces duplicate material spaces in various functions such as retailing, consumption of goods and services (with entertainment being an important component), political and community organizing, as well as business and peer-to-peer communication and social networking. Zook (2001) demonstrated that online contents closely follow the geography of contents producers, and Aoyama and Sheppard (2003) argued that real and virtual spaces are dialectically linked, in that geographic space offers defining principles that profoundly shape the structures of virtual spaces. Geographic context, territorial structures of regulation and physical communications infrastructure shape, to a large extent, production and consumption in cyberspace, and geographically induced frictions play a role in the relationship between virtual and real places. In sum, as advocated by Leamer and Storper (2001), the Internet will work to reinforce urbanization on the one hand, and on the other hand, allow dispersion of certain activities.

197

KEY POINTS

- Knowledge economy is closely associated with economic sector theory and the emergence of the service economy, and was based on the observation that the knowledge workers were gaining significance in advanced industrialized economies.
- The Internet, on the one hand, amplified the importance of knowledge work in the new economy, yet on the other hand, created new possibilities in organizing economies.
- The geography of the knowledge economy suggests spatial concentrations of innovation, as knowledge workers tend to live and work in certain cities and regions.

FURTHER READING

For a recent work on the role of knowledge in economic geography, see Gertler and Levitte (2005). For research on the creative class in Europe, see Lorenzen and Andersen (2009). For a more recent research on the geography of cyberspace, see Zook and Graham (2007).

NOTES

1 Pecuniary external economies is one type of externality that affects others through the market (i.e. prices).
2 Exportable services may include, for example, accounting services, financial services and any other types of services for which a consumer may be located at a significant distance from the location of firms.

6.2 FINANCIALIZATION

The total daily turnover of international financial markets grew exponentially in the latter part of the twentieth century. Also, the shares of total employment and GDP in the financial industry have been steadily increasing in many economies. Despite the importance of financialization, no consensus has emerged on its definition. To some, it refers to the growing importance of profits generated by speculative investment. Some observers focus on the greater role of finance in individual, household, and firm behaviour (Erturk et al., 2007), while others consider the role of finance in causing systemic changes in capitalism (Strange, 1986; Boyer, 2000). Yet others view financialization as the continuous evaluation of economic investments by the financial markets.

Origins and fundamentals

The emergence of money as a medium of exchange enabled the development of the market economy and of international trade. Money not only allows exchange, but also has led to a number of financial innovations. One decisive innovation was the inception of account money (i.e. accumulations of commercial credits or household savings, creation of debt, and generating liquidity from previously fixed assets), which led to the development of banking and other financial institutions. In the seventeenth century, various financial markets were established, including equity (stock), currency, options and futures markets. The currency regime has undergone transformation, from the previously dominant silver standard to the gold standard; the latter became officially adopted by Britain and the United States in the nineteenth century. The two world wars made the gold standard difficult to sustain, however, and in the post-World War II period the Bretton Woods Accord (1944–71) established the US dollar as gold convertible with other currencies pegged to the dollar, along with the establishment of two key international financial institutions, the World Bank and the International Monetary Fund. Under the temporal stability of currency exchange, international financial transactions grew, and an increasing 'deterriotorialisation of credit provision' (Leyshon and Thrift,

1997) took place, most notably characterized by currencies traded outside their home territories such as the Euro-dollar (i.e. US dollar traded in Europe to profit from higher interest rates).

Financial markets can be categorized into primary and secondary markets. In primary markets, new financial assets are raised for corporations. Typically, this is done by intermediaries (such as banks) that issue, price and sell the securities, i.e. corporations' equity or debt, to investors. These new financial assets are traded before maturity in the secondary markets, such as stock exchanges, between investors. Financial markets can also be divided into over-the-counter markets (direct exchange between two parties) and organized exchanges. Organized exchanges standardize terms of transactions, and therefore lower entry-barriers and risks for investors, whereas non-standardized products are mostly traded in over-the-counter (OTC) markets. Financial markets can be further classified into those where various securitized commodities are traded (e.g. equity markets, commodity markets, commercial real estate and residential mortgage markets), and those that handle financial transactions without direct links to commodities, such as bond, currency, money and derivatives markets. Some of the financial markets mentioned above can be highly volatile, in particular equity and currency markets.

In economic geography, the study of finance began as one way to address the 'productionist bias' of the discipline, which was most prevalent in the 1970s and 1980s. Today, financial geography is an active component of the discipline and ranges from the study of global cities to forms of governance in the pension fund industry. One important geographical aspect of the financial economy concerns the process of transforming real assets into financial liquidity, which in turn makes capital highly mobile. Furthermore, demand for returns in finance is uncharacteristically short term when compared with real investment. The geographical hyper-mobility and short-term nature of finance has in turn produced a speculative global economy. The mobility of capital has increased the disconnection of money from the real contexts of the local economy.

Regulation in finance

Financial markets have direct links to government policies – particularly with monetary policy – which is typically governed by central banks.[1]

The formation of the European Monetary Union represented an important regulatory change that was instrumental in implementing the single European market and emerged as a strategy to combine forces to more effectively counteract the dominant currencies such as the US dollar and the Japanese yen and their respective monetary systems. Financial markets are subject to numerous regulations with basically two general aims, to prevent systemic crises and to protect retail investors and bank clients.

The history of finance is characterized by progressive and large-scale de-regulation of cross-border controls on financial flows, particularly in the last few decades (Laulajainen, 2003). In the United States, progressive deregulation since the mid-1970s allowed for more competition by, for example, removing deposit interest rate ceilings and restrictions for competition; allowing new product introduction and inter-state banking. Deregulation is claimed to have led to the growing failure of commercial banks that peaked in 1990 (Warf and Cox, 1995). Progressive deregulations that minimized government involvement are viewed to have accelerated the process of financialization. Governments also serve as important guarantors of financial instruments (such as by issuing money), but fraud and taking advantage of loopholes in the regulations still occur. Strange (1986) termed the contemporary financial system 'the global casino of high finance' (or casino capitalism), and was concerned with the grave consequences that would impact everyone, including farmers, shop-keepers and factory workers. Particularly since the 1970s, volatility in currency markets and interest rates have made various economies vulnerable to financial crisis.

201

Speculative and fictitious capital

Marx (1894) called financial capital *fictitious*, since it represents an advance claim upon the yet-to-be produced value, and does not generate value until it is transformed into real investment. The development of secondary markets mentioned above has played a major role in making financial capital fictitious. Financial assets are immaterial, abstract and fundamentally implicit promises, which therefore inherently involve risk (Laulajainen, 2003). Financial risk reflects the variability in rate of return, and investors typically respond to risk by diversifying their investments. Professional investors, typically called fund managers work on behalf of institutional investors (the largest of which are investment banks, insurance firms and pension funds); such investors not only acquire, trade, arbitrage[2], insure and hedge against (cross-insure) various types of risks, but also engage in trading highly speculative financial instruments, such as insurances and derivatives, diversifying risk further. Speculation is integral to the functioning of financial markets (Clark, 1998).

The financial economy is also contrasted with the 'real' economy, one that relies on production and delivery of tangible goods and services for economic growth. The financial economy centres around managing portfolios of assets (Markowitz, 1959). Financial assets are typically measured by equity (i.e. ownership of property or productive capacity such as firms), debt (i.e. loans), and bank deposits (i.e. corporate as well as household savings). Financial markets have facilitated the formation of global circuits of capital (see 4.3 Circuits of Capital) which enable investment to move across various economies.

Finance in economic development

Financing start-up firms requires risk capital that can endure high risks for the potential of high returns. Venture capitalists and business angels[3] are key financiers of start-up firms, thereby playing a critical role in supporting entrepreneurship and innovation (see 2.2 Entrepreneurship; 2.1 Innovation). The venture capital market, which emerged first in the United States in the late 1940s, has been instrumental in promoting active post-war entrepreneurship (see Leinbach and Amrhein, 1987; Florida and Kenney, 1988). In the dot-com boom of the late 1990s, major venture capital firms around Silicon Valley played a significant role in providing initial capital for many start-ups (Zook, 2004). Similarly, in the Global South, the Grameen Bank in Bangladesh and other micro-finance institutions offer micro-credit to low-income people, mostly to women as a way to offer day-to-day survival through modest financial independence.

Interests in finance are also represented by (partly government sponsored) the increasing importance of pension funds as investors in the economy. Deregulation in the form of pension restructuring (Clark et al., 2001; Engelen, 2003; Clark and Wójcik, 2005) has meant a greater involvement of pension funds as investors, a process motivated by the promise of higher returns, yet it also exposes pension funds to increasing risks. Clark (2000) refers to the significance in size and role of pension funds as distinguishing features of Anglo-American financial markets (Pension Fund Capitalism). He also emphasizes their emerging role as investors in public assets such as urban infrastructure and community economic development projects. In addition, longer life-expectancies, changing demographics and shifting expectations towards life after retirement contribute to the increasingly complex challenge faced by pension funds in balancing risks with returns in financial markets.

Households and individuals

Individuals and households have increased their involvement in financial markets in recent decades. Financial innovations such as consumer credit (e.g. personal credit cards) were conceived to encourage and further promote the act of household consumption (see 6.3 Consumption). Although the primary clients of financial markets are institutional investors, such as banks, insurance companies, pension funds and corporations, individual investors are increasingly involved both directly and indirectly.

A series of deregulations and widespread access to the Internet have allowed individuals to engage in the buying and selling of stocks, currencies, and futures (i.e. online traders). Until the financial crisis of 2008, individual middle-class online traders, including Japanese housewives, were actively trading various financial products on the market (Fackler, 2007; McLaughlin, 2008). Furthermore, household savings today are mostly grouped into mutual funds and pension funds. The dependence on pension funds means that the economic well-being of households is increasingly influenced by price fluctuations in the financial markets (see Langley, 2006).

Financial crises

Financial crises involve failing banks. Although related, financial crises can be broadly categorized into two causes, one arising out of a solvency crisis (liabilities exceed assets) and the other arising out of a liquidity crisis (on-time payment of bills). When financial crises occur, typically the lender of the last resort (usually the government) is called in for a rescue. In the United States, commercial bank failures grew dramatically in the 1980s, prompted by a nationwide commercial real-estate glut and a petroleum-glut-induced recession in Texas (Warf and Cox, 1995). This resulted in the US government bailing out many savings-and-loan institutions, and partially caused the government deficit in the 1990s. Each past crisis has resulted in regulatory reforms, yet such reforms do not seem able to avert the next financial crisis. The government bail-out is controversial not only because it spends tax-payers' money for bad business decisions, but also because it creates an expectation in the business community that the government will ultimately step in, thereby encouraging, rather than discouraging, reckless financial decisions.

Countries can also experience financial crises. The Latin American Debt Crises included Mexico, Argentina and Brazil which, in the early 1980s, became unable to service (repay interest of) their external debts. These loans were typically made to finance infrastructural projects which were designed to support import-substitution industrialization policy. Many of these loans, however, were short-term loans, which required frequent refinancing. Once the crisis became apparent, refinancing became difficult, which prompted involvement of the IMF as well as debt relief and rescheduling programs. Leyshon and Thrift (1997) argued that US government policy of sustaining high interest rates during those times in order to dampen domestic inflation in part contributed to these crises.

The Asian Economic Crisis (1997–98) and the global financial crisis of 2008 were initiated by escalating commercial and residential real estate prices. In the case of the former, hyper-growth had prompted aggressive commercial real-estate speculation in the greater Bangkok metropolitan area, and in the case of the latter, the sluggish performance of the post-dot-com boom US equity market, combined with historically low interest rates had channelled financial capital to the housing mortgage market, leading to the proliferation of sub-prime mortgage loans, which precipitated the crisis. The fall of investment bank Lehman Brothers to bankruptcy protection and the purchase of Merrill Lynch by Bank of America made the stock market tumble worldwide on 15 September 2008, prompting a system-wide liquidity crisis.

Some argue that the increasing complexity in financial tools (and the inability of regulatory agencies to keep up with them) in part contributed to the global financial crisis of 2008. Multiple layers of trading compound the complexity and opacity of the financial system. The contemporary financial sector employs a vast number of mathematicians whose intricate models and calculations drive much financial innovation and make what is already a complex system even more opaque and difficult for regulators to comprehend. Future research in this area calls for economic geographers to study the geographic sources as well as manifestations of financial crises, as well as geographic variations in the performance of banks.

The geography of finance

In spite of the claim that financial integration has led to the end of geography (O'Brien, 1992), there are significant variations in the geography of financial markets. For example, whereas currency and bond markets are truly global in that they can be traded anywhere, equity still tends to be traded close to the headquarters of firms (Laulajainen, 2003). Both concentration and dispersion characterize the geography of finance. On the one hand, financial actors are to a large extent concentrated in global cities such as New York (Kindleberger, 1974; Friedman, 1986; Sassen, 1991), London (Pryke, 1991, 1994; Thrift, 1994) and Tokyo (Rimmer, 1986; Sassen 1991), which exert considerable influence on the world economy. The financial sector is among the most sensitive to the quality of information, thereby making face-to-face contacts particularly important (Agnes, 2000). On the other hand, financial firms were among the most aggressive in adopting information technology (IT), which not only sustains second-tier financial centres such as Frankfurt (Grote et al., 2002), Amsterdam (Engelen, 2007), Dubai, Hong Kong and Singapore, but also led to the development of offshore financial centres in the periphery to avoid regulations. Some of the offshore financial centres emerged in places that were already tax havens, such as the Cayman Islands, the Bahamas, Luxembourg, Liechtenstein, and Bahrain (see Roberts, 1995).

Financialization is also the process of integrating local economies into the global economy (Tickell, 2000). The centralization of finance in major cities exacerbates the decline of local banking systems in peripheral areas, which might have repercussions for the ability of small local firms

205

to raise necessary capital (Dow and Rodriguez-Fuentes, 1997) (see 5.5 Networks). Within these financial centres, the separation between front-offices in prime downtown locations and back-offices in suburban locations became increasingly prevalent in the United States during the 1970s (see Nelson, 1986; Dicken, 1998) (see 6.1 Knowledge Economy).

Yet, global convergence has hardly been achieved in the financial markets. Traders around the world are highly heterogeneous in their backgrounds, culture and reactions (Agnes, 2000; Thrift, 2000). For example, Pollard and Samers (2007) have shown how Islamic law and associated cultural practices shape risk perceptions and result in distinctive Islamic financial institutions and products. Traders also make assumptions on the stability and credibility of various financial markets based on their limited knowledge. During the Asian economic crisis that began in 1997 in Thailand, traders not only pulled out of Thailand but also from markets in neighbouring economies simply by the virtue of proximity and geographical associations. Most recently, interests in alternative finance – such as local currency movements (Lee, 1996) and ethical investing – have grown as a way to inject a moral economy and social choice into the financialized global economy.

206

KEY POINTS

- Although no consensus exists on the definition of financialization, there is no doubt that the financial economy has grown in its size and significance in the world economy, with some claiming that it indicates a systemic change in capitalism.
- Regulations play a central role in financialization, and can promote economic growth by inducing innovation through risk capital financing and creating investment opportunities for households, but regulations also can also become the cause of financial crises.
- The importance of geographical proximity in the financial industry gave rise to financial centres, which contribute to industry specialization as well as uneven territorial development.

FURTHER READING

For an overview of the financial crisis of 2008, see Dore (2008) and Ashton (2009) for the history of the subprime mortgage market in the

United States. A section in the Bank for International Settlements 2008 Annual Report also explains the financial crisis in an accessible manner. In addition, a National Public Radio programme, *This American Life*, featured the mortgage crisis in episode 355: 'The Giant Pool of Money' (9 May 2008). The transcript is available at: http://www. thisamericanlife.org/Radio_Episode.aspx?episode=355.

NOTES

1 In the United States, the Federal Reserve System ('The Fed' for short, established in 1913) serves as the central bank, and the Board of Governors of the Federal Reserve System, located in Washington, DC, sets monetary policies.

2 Arbitrage refers to an act of simultaneously buying and selling financial commodities such as securities, bonds, currencies or futures contracts in more than two markets in order to take advantage of price differentials between markets.

3 Business angels are wealthy individuals, who play the role of investors, providing initial capital to those entrepreneurs with promising business ideas without formal market-based transactions.

6.3 CONSUMPTION

Consumption is an act of spending resources to fulfil basic needs (e.g. food) as well as a way of expressing social status, tastes and preferences. While the latter is of particular interest to sociologists and cultural geographers, economic geographers' interest in consumption began with a focus on its role in shaping economic landscapes of human settlement. This interest prompted locational and organizational studies of the retail sector, a service sector directly engaged with consumption. Although consumption has long been secondary to production as a central focus of economic geography (known as 'productionist-bias' (Wrigley et al., 2002)), recently there is a renewed interest in the geographic manifestations of a mass-consumer society, including consumer activism.

Location theories of consumption

The locational characteristics of the retail sector have been studied extensively since the early twentieth century, and many of the basic principles developed in these theories remain broadly applicable today. Classical location theorists such as Harold Hotelling (1929) and Walter Christaller (1966[1933]) theorized the spatial outcomes of competition and the optimal allocation of retail services to serve spatially dispersed populations. Hotelling's model considers how retail competition stabilises geographically under 'duopoly', a condition where two competing merchants are assumed to be located along a two-dimensional line, selling an identical product. Popularized by the analogy of two ice-cream salesmen moving along a beachfront to find optimal locations, Hotelling showed that each vendor will end up locating back-to-back at the centre of the beach in order to maximize his/her sales area and prevent the competitor from capturing a larger area. Hotelling's analysis explains why retail functions agglomerate rather than disperse across space, and through this example, offered an explanation for a relatively neglected phenomenon in economic theory at the time, a condition called geographic monopoly. This model, however, did not take into account demand elasticity – the consumer's propensity to purchase a certain quantity at a certain price. From the consumer's perspective, the price of a product includes travel cost (getting to a store from home), which complicates the reality of retail location.

Walter Christaller (1966[1933]) developed a location theory that partially addressed the problem of demand elasticity by incorporating the consumer's travel costs. Christaller's central place theory was derived from the observed hexagonal patterns of the settlements in Southern Germany (his theory was later adapted for Southwest Iowa in the US by Berry, 1967). His two principles, the range of a good and the threshold of a good, refer respectively to the maximum distance that consumers are willing to travel to purchase a good/service, and the minimum market size needed for retailers to stay in business. Integrating these two principles, an optimal spatial pattern emerges, characterized by hexagonal market areas and a hierarchy of different-sized settlements around the central place of each hexagonal territory. Christaller conceptualized that these central places would facilitate efficiency through the development of a small number of higher-order centres (larger cities) and a larger number of lower-order centres (smaller cities and towns). Higher-order centres are places where high-order goods (high-priced items purchased infrequently, such as cars or jewellery) are traded, whereas lower-order centres are more easily accessed through reduced travel costs, and serve as places where low-order goods (low-priced items purchased frequently, such as milk and eggs) are traded. Christaller's model, however, is built on a set of simplifying assumptions, including a uniformly distributed population of equal income living on a flat surface without physical barriers (such as mountains that make travel costly), a single mode of transport, and travel costs that are proportional to distance. Consumers are assumed to be economically rational individuals who minimize costs and maximise resources. The failure of the real world to adhere to these assumptions distorts the hexagonal patterns posited by Central Place Theory (Berry 1967).

209

From retailing to buyer-driven commodity chains

Hotelling and Christaller's models laid important foundations for understanding retail location. The principles of threshold and range still retain relevance. From the consumption standpoint, consumers minimize travel distance depending on the goods/services in demand.

For example, consumers typically minimize travel distance to purchase daily necessities such as milk and eggs and are willing to travel longer distances to buy jewellery for a mother's 60th birthday. However, except for a few case studies on consumer choice (see Clarke et al., 2006; Wang and Lo, 2007), contemporary economic geography research in consumption has shifted its focus from the locational characteristics of consumption to the organizational characteristics of the retail sector, the sector that most directly engages with the consumer.

Today, economic geographers are broadly concerned with two major aspects of retail restructuring. One is the dominance of large-scale retailers at the expense of small retailers, and the other is the dominance of foreign retailers over domestic retailers. These concerns evoke particularly strong emotional responses among consumers and shape political agendas for consumer activism. Concerns about the dominance of large-scale retailers are as old as the emergence of department stores in cities such as New York, London, Paris and Berlin in the latter part of the nineteenth century (see, for example, Gellately, 1974). Similarly, the proliferation of successful chain stores and supermarkets originating in the United States beginning in the 1930s, such as A&P, Sears and J.C. Penney, has raised concerns about the fate of small retailers. While earlier theories considered small retailers (e.g. 'mom-and-pop' stores) archaic, inefficient and in need of modernization (Goldman, 1991), others view them as an important source of personalized service, entrepreneurship, community identity and job opportunities for the working classes, and a counterweight to the monopolistic tendencies that reduces incentives for innovation.

Concerns about the dominance of foreign retailers became prominent with the emergence of Retail MNCs in the latter 1990s such as Wal-Mart (USA), Carrefour (France) and Tesco (UK). Retail MNCs emerged as a result of push-factors (market saturation among others) from the home market as well as pull-factors (deregulations in the host markets providing new market opportunities). A number of retailers have globalized their production through buyer-driven commodity chains (see 4.4 Global Value Chains). Typically, large retailers have dedicated subcontractors in low-income economies to cater to consumers in high-income economies. Some have successfully developed a highly sophisticated and demand-response global logistics system to deliver affordable products that cater to fast changing fashion trends (such as the Spanish apparel store chain Zara). Yet, many retailers still conduct research on store design and marketing in a few select cities (such as Paris, New York, Milan, Toronto and San Francisco) and actively engage in brand marketing to promote product differentiation and customer loyalty.

Mass-consumer society and consumer behaviour

Over the past two centuries, people in high-income countries have experienced a gradual decrease of work-time, and the relative expansion of non-work time. As noted first by Veblen (1899), this shift has prompted lifestyle changes that have shaped consumption behaviour, particularly growth in the conspicuous consumption of leisure, entertainment and tourism products/services. Max Weber viewed class and status as interrelated yet distinctive categories, in which class was related to household production (income generation), while status was related to consumption (Gerth and Mills, 1946).

Under high mass consumption, the most advanced stage of economic development according to Rostow (1953), leading sectors employ highly skilled workers to produce durable consumer goods and services, workers and their families have high levels of disposable income to spend on more than just basic food, shelter and clothing, and various technological and household gadgets become widely adopted. In many mass-consumer societies, consumption has become the primary driver of urban development and re-development, prompting processes of gentrification, converting earlier production landscapes such as harbours and factories into consumption landscapes such as waterfront commercial districts and residential lofts (see Zukin, 1995).

A notable trend in contemporary mass-consumer societies is the globalization of tastes. On the one hand, homogenization of cultural tastes is observed, as represented by the global popularity of Levi's jeans or rap music. On the other hand, some argue that the emphasis is shifting from particularity of tastes (in 'high-art' such as opera) to diversity of tastes (e.g. world music). In high-income countries, 'spending fatigue'[1] combined with an increasing cultural diversity brought on by international immigration are contributing to a greater curiosity and demand for products that are 'foreign' and 'exotic'. The popularity of world music, for example, represents openness to and respect for the cultural

211

expression of others, yet it is packaged according to the cultural tastes of the privileged (Connell and Gibson, 2004). The diversity of tastes has emerged as an important aspect of class distinction, and motivates consumers to choose leisure and entertainment products, including tourism destinations.

Class distinctions have geographic implications, as they motivate consumers' choice of leisure and entertainment products and tourism destinations. For example, tourist destinations have diversified to increasingly off-the-beaten-track locations. Also, tourism used to be practiced exclusively by those from the advanced industrialized societies, but in the past decade tourists have become more diverse, with a growing number of tourists from Eastern Europe, Russia and Asia). Policy makers have noticed the growth of tourism and the increasing diversity of niches in tourism (i.e. eco-tourism, cultural tourism) (see 5.1 Culture). As a result, tourism has become an important economic development strategy, particularly in the developing world where the skills, resources and technologies needed for manufacturing activities are in short supply (see Kozul-Wright and Stanbury, 1998). Opinions on consumption-driven economic development models remain divided, however. Some consider production too important to be ignored, and others question the conceptual validity of the consumer-driven economy (see 6.1 Knowledge Economy).

Spaces of consumption

Regulation plays an important role in shaping various aspects of consumption, through planning and zoning and store-size laws (to protect residential neighbourhoods and manage urban sprawl), competition laws (typically in the form of anti-trust regulations to protect small retailers from over-competition and regulate fraudulent practices) and store opening hours (to protect workers from exploitation). In general, stricter retail regulations tend to favour small retailers. For example, in Europe the influence of regulation on consumption is especially important as the state has assumed a stronger interventionist role than elsewhere (Marsden and Wrigley, 1995, 1996). Also, cultural factors, reflected in shopping habits (daily versus weekly grocery shopping), cultural preferences and diet (fresh versus processed food), the gender division of labour in household chores and available modes of transportation (public transit versus automobiles) all shape the spatial and organizational structures of the retail sector. For example, the continued popularity of traditional 'wet-markets'[2] in East and Southeast Asia, despite the introduction of supermarkets, suggests that cultural factors are important elements shaping cross-national differences in the retail sector (Goldman, 2001).

Economic geographers today receive inspirations for research on consumption from social and cultural geographers (see Crang, 1994; Goss, 2004, 2006) and from sociologists (for example, Zukin, 1995; Bhachu, 2004). For example, structural social inequalities based on gender, class and race are an essential aspect of consumption, and may lead to social exclusion of various types (see Domosh, 1996; Gregson and Crewe, 1997; Williams, 2002) (see 5.2 Gender).

Contemporary consumer cultures fetishize commodities, refashion identities, commodify differences and thereby produce distinctive symbolic geographies (Jackson, 2004). Sayer (2003) takes a 'moral economy' approach which brings in issues of subjectivity and normative dimensions by invoking such concepts as 'sign value' or association of meanings by users in analysing the commodification of culture. Others have adopted the global commodity chain framework to examine the role of culture and commodity. Leslie and Reimer (2003), for example, analyzed the intersections between fashion and furniture industries and their manifestations in magazine, retail and manufacturing spaces.

213

Economic geographers are also joining cultural geographers to study, for example, the popularity of second-hand stores and local-currency experiments (see Gregson and Crewe, 2003; Leyshon, 2004). A growing field of geographic research in recent years has also explored the emergence of alternative spaces of consumption. Work on alternative exchange networks emphasizes how socially anchored and locally defined networks of consumption, exchange and re-use, are important components of regional economies (see Gibson-Graham, 1996).

Consumers are increasingly recognized as an important source of creativity. Today, the consumer's role is far greater than what is traditionally considered either as passive or at best 'consultative'. Rather than being situated simply as an end-user who exercises purchasing decisions, consumers are increasingly viewed as co-producers, not only of software codes but also of various cultural contents, including fashion, music, video games and films (distributed via the Internet through various websites such as MySpace and YouTube). The boundary between the consumer and producer has been redefined in a new way with the onset of the Internet, further blurring the distinction between the producer and the consumer (von Hippel, 2001) (see 2.1 Innovation). Consumers not only benefit from customisation, but they also increasingly derive satisfaction and gratification, and even achieve social status within a certain community, by showcasing their talent.

Power of the consumer

The consumer's role in the economy has been dramatically transformed during the past century (Larner, 1997). With the diminished power of labour unions, individuals and groups have turned to purchasing power as a primary means of influencing corporate behaviour and advocating for consumer well-being.

Consumer activism is a major way through which individuals and groups influence the course of capitalist development. As production has globalized and often moved into areas viewed as having no or inadequate labour and environmental regulatory mechanisms, conventional regulatory mechanisms, which are territorially based and largely intranational in scale, are no longer able to control the global commodity chain. Through consumer activism, individuals and groups seek to improve labour conditions with long work hours, low wages, an unsafe work environment, inadequate protection of workers and human rights. Two areas where consumer activism has been particularly prominent are movements against sweatshops (most frequently in the garment, apparel and shoe industry) and movements to protect agricultural workers in developing countries who often face increasingly volatile price fluctuations of the global market (such as coffee), and/or exposed to dangerous levels of pesticides and chemical fertilizers. Some believe consumer campaigns and boycotts against corporations and brands are instrumental in ensuring corporate social responsibility while others believe these activities hurt farmers and factory workers in developing countries by pressing down their wages and subjecting them to dangerous working conditions.

Consumers are playing a more active role in the way we use various types of resources. The increased distance between production and consumption is also observed in resource-based industries, including agriculture, agro-food industries, forestry and petroleum industries (see Hughes and Reimer, 2004). Environmental Defense Fund, a US-based non-profit advocacy group, for example, produces a 'pocket seafood selector' which ranks seafood by such criteria as mercury contents and method of catch, so that consumers can make better choices about what they eat at home and at restaurants. Although the organic farming movement goes back to the environmental movement against the use of chemical fertilizers in agriculture, in the 1990s and 2000s it became combined with consumer activism primarily concerned with food safety (see Hughes et al., 2007). The slow food movement supports local farmers to promote locally grown food through co-op movements or to provide incentives for safe and environmentally sustainable practices such as organic farming. These types of consumer activism seek to increase the transparency of food production as supply chains expand globally, and to ensure that agricultural workers in developing countries are receiving a living wage (fair-trade movements) (see Hughes and Reimer, 2004). Today, organic food has become part of the mainstream, with new niche retailers such as Whole Foods (USA) and Waitrose (UK), as well as large retailers like Wal-Mart or Tesco, recognizing the revenue potential from the organics.

Another emerging trend is the rise of cyber-activism. Whereas the Internet has undoubtedly facilitated various consumer groups to organize and initiate campaigns and boycotts, the Internet has also allowed individuals to broadcast negative experiences with corporations. This ranges from a disgruntled customer unhappy with the responses of the customer-service division of a laptop he purchased, to the proliferation of various consumer-rating websites for online merchants, hotels, restaurants, entertainment products (such as music, books and game software) and travel destinations.

215

KEY POINTS

- Consumption has traditionally been understood through locations and organizations of the retail sector.
- Globalisation and the extensive reliance on global commodity chains have fundamentally altered the organisation of the retail sector and choices for consumers.
- The view on the role of consumption in the economy is changing from one of passive to active involvement, through consumer activism, co-production, and user-led innovation.

FURTHER READING

For a discussion on the role of the consumer in innovation, see Grabher et al. (2008). For a discussion on the contemporary issues on the retail MNCs in economic geography, see Coe and Wrigley (2007).

NOTES

1 It has been suggested that market saturation and abundance has made consumers bored of consumption, making them seek out unique 'experiences' in consumption (Pine and Gilmore, 1999).

2 Traditional marketplaces observed in parts of East and Southeast Asia, where vegetables, meat and seafood (including live animals such as chicken and turtles) are sold by individual vendors.

6.4 SUSTAINABLE DEVELOPMENT

The concept of sustainable development integrates ecological, social and economic concerns into a single developmental model that strives to ensure inter-generational equity with respect to material, social and environmental well-being. Although the concept has been widely critiqued, it has inspired economic geographers to more carefully and critically examine how economic activities affect environments at the local and global scale. Three perspectives on sustainability have marked economic geography's engagement with the concept: how industries can be made more sustainable; how communities can more effectively practice sustainable development; and how global climate change is increasing the socio-economic vulnerability of people and places throughout the world economy.

Origins and fundamentals

The origins of sustainable development have been well documented by a number of scholars (e.g. Dresner, 2008; Adams, 2009) who trace the concept back to early environmental activists and scientists such as Aldo Leopold, Rachel Carson, Paul Ehrlich, Donella Meadows and E.F. Schumacher. These thinkers observed the pace, scale and consequences of industrialization and mass consumerism and argued that humanity's long-term survival was only possible if global society more effectively balanced its material relationships with nature by reducing consumption, cleaning up industries and limiting population growth. By the 1980s, sustainable development had gained a foothold in mainstream governments and development institutions and the concept was first formally defined in the World Commission on Environment and Development's (WCED) Brundtland Report, *Our Common Future* (WCED, 1987).

Sustainable development as defined by the WCED means that societies must balance the desire to maximize the benefits of economic and industrial growth with a need to maintain the quality of, and services from, natural resources and ecosystems. The WCED definition emphasized three ideas about resource use and socio-economic equity. First, renewable resources (e.g. solar power, animal and plant materials) should be consumed only at rates lower than their natural rates of regeneration and renewal. Second, non-renewable resources (e.g. oil, minerals) should be used efficiently toward an optimal short-term social good but with the understanding that substitute resources will be created through technological change or new discoveries. Third, inter-generational equity must be a guiding principle for sustainable development policies. This means that meeting the material needs of current generations must not be achieved at the expense of future generations' abilities to have equal or greater opportunities. To achieve these goals, the WCED argued that all nations must contribute to sustainability and that this could be achieved through multinational institutions and commitments to global treaties on environmental protection (e.g. the Global Environment Facility, the Kyoto Protocol).

The WCED definition is regarded by most as a weak form of sustainable development. Proponents of weak sustainability argue that economic growth can have a benign effect on the environment if renewable resources are used as much as possible, environmentally friendly technologies are developed and if humanity maintains a minimum or survival level of capital in all its forms (e.g. human capital, natural capital and physical capital) (World Bank, 2003). Market forces are best suited to drive the transition toward a sustainable society and some proponents of weak sustainability argue that a short-run focus on economic growth will lead to a long-run phenomenon known as the Environmental Kuznets Curve, a situation where increases in per-capita income are paralleled by declining pollution levels (Grossman and Krueger, 1995). Weak sustainability has been promoted by development agencies such as the World Bank (2003) and it is commonly associated with the field of environmental economics where neo-classical economic models are applied to environmental issues and problems (e.g. see Tietenberg, 2006). Eliminating negative externalities such as pollution, land degradation and species extinction are common areas of research and environmental economists emphasize the role that markets can play in preventing these. One key idea is the polluter-pays principle, i.e. the notion that cleaner industries will develop if states force firms to pay for all the social and environment costs associated with the pollution they create.

Weak sustainability arguments have been criticized for their failure to address the fundamental constraints that natural resources, energy flows and ecosystem services place on neo-classical models for economic development. Proponents of strong sustainability adhere to a more Malthusian position arguing that unending growth in population and resource consumption will lead to dire ecological and socioeconomic consequences.[1] There are limits to growth that natural resource substitution and technological changes cannot overcome and the planet will be better served by moving toward a steady-state economy, one that balances human survival and biogeophysical systems within the boundaries determined by the Earth's carrying capacity (Daly, 1977; Daily and Ehrlich, 1992). The field of ecological economics was developed to model this alternative economic system and its scholars view the Earth's natural and human systems (e.g. the atmosphere, hydrosphere, biosphere and the global economy) as interconnected and interdependent on one another (Daly and Farley, 2003). The interconnectivities and interdependencies mean that perpetual growth of the economic system can be achieved only through potentially irreversible damage to other planetary life-support systems. As such, development policies should be guided by environmental science and the precautionary principle must be applied to all policy decisions, i.e. the notion that actions to prevent irreversible damage to Earth must preclude scientific certainty about an issue's causes or consequences (e.g. carbon emissions must be reduced immediately even if there is uncertainty about the long-run implications of global warming).

219

Critiques of sustainable development

Although sustainable development has received support in a wide range of academic fields, the concept has also been subject to numerous and significant criticisms. For Redclift (1993, 2005), sustainable development is oxymoronic, ambiguous and essentially a new form of economic modernization that leaves intact the neo-classical paradigm of unending growth. Particularly problematic is the issue of what constitutes a basic human right or need since the definition depends on who decides, where they live, and on the assumptions they make about the desired needs and rights of future generations. Related criticisms centre on the concern that the true meaning and spirit of the concept has been corrupted by institutions (e.g. the World Bank) that prioritize economic objectives over social equality and environmental protection

(Sneddon et al., 2006). Other critiques are more forceful, arguing that sustainable development is simply a discursive tool that will enable the continued expansion of destructive and exploitative forms of capitalism (Goldman, 2004).

While there are numerous alternative visions for human–environment relations (e.g. deep ecology, eco-socialism and eco-feminism) (see Adams, 2009 for a discussion) – geographers have generally confronted the sustainability concept through the field of political ecology (e.g. Robbins, 2004). For political ecologists, the implications of, and possibilities for, sustainable development can be fully understood only when the concept and its policies are situated in relation to the political-economic institutions (e.g. states, class relations and development agencies) from where it originated. When considered this way, sustainable development is no longer a scientific theory but a political project created by a group of elite individuals who come from particular places and who have a vested interest in retaining and/or strengthening their socioeconomic power (Bryant, 1998). In applying these ideas, political ecologists have highlighted some of the contradictions and negative consequences of sustainable development initiatives (e.g. McGregor, 2004).

220

Economic geography and sustainable development

Interest in sustainable development has helped to create the subfield of environmental economic geography. Scholars in this area strive to understand how the spatial and geographical characteristics of economic activities influence environmental quality and shape the prospects for more sustainable communities, industries and regions. Most research focuses on one of three themes: the sustainability of industries; the sustainability of urban and rural livelihoods; and the links between global climate change, economic globalization and socio-economic vulnerability.[2]

Research on the evolution of industrial systems focuses on how country or region-specific factors and institutions shape sustainability trajectories. Studies of industrial transformations in Asia have been particularly significant and have demonstrated the importance of effective governance, policy formulation and environmental monitoring for sustainable development (Angel and Rock, 2003; Rock and Angel, 2006).

Sustainability transitions are best achieved when firms develop the capabilities needed to adopt global best practices with respect to environmental performance and when local regulatory agencies are respected, autonomous and able to closely monitor industrial-pollution levels. A related area of research draws on ideas from industrial ecology and ecological modernization (see Frosch, 1995; Huber, 2000) and links these to theories on industrial clusters and districts. Eco-industrial clusters – agglomerations of companies where one firm's waste is another's input (e.g. BASF's *verbund* chemical complex in Germany) – can reduce the environmental impacts of industrial growth provided they overcome the numerous challenges to their successful development (Gibbs, 2003; McManus and Gibbs, 2008) (see 3.2 Industrial Clusters).

Beyond the firm and regional scale, the sustainability of industries also depends on how global trade flows and post-consumption waste are managed and regulated. For example, the amount of electronic waste or e-waste derived from discarded electronic devices (e.g. computers, mobile phones and audio equipment) has grown exponentially since the 1970s and a global e-waste recycling industry has emerged, one that has had particularly negative impacts on human health and the environment in the Global South (Pellow, 2007). To counteract this trend, business leaders, policy makers and environmental activists are developing global standards and regulations for the management and trade of e-waste (e.g. see the Basel Action Network webpage, www.ban.org).

221

Economic geographers have also studied the prospects for more sustainable cities and rural communities. Cities create large amounts of waste and pollution through the consumption, production and distribution activities in them and thus have very large ecological footprints – demands for resources (i.e. energy, food, forest products and the built environment) that are measured in land area (Wackernagel and Rees, 1996). Researchers have examined how ideas about sustainability are being integrated into planning strategies and urban brownfield (i.e. former industrial sites) redevelopment initiatives (McCarthy, 2002; Counsell and Haughton, 2006), and about what these initiatives mean for social and environmental justice issues (Agyeman and Evans, 2004; Pacione, 2007). Housing is also an important theme both in terms of how new housing developments can create unsustainable 'consumption landscapes' (e.g. Leichenko and Solecki, 2005) and how changes in household consumption and waste management practices influence the sustainability of cities and suburbs (e.g. Barr and Gilg, 2006). With respect to rural communities, geographers have examined the prospects for community supported agriculture and 'alternative' food networks in industrial countries (e.g. Maxey, 2006) and how the sustainability of rural livelihoods in the Global South has been challenged by economic liberalization programmes (e.g. Bebbington and Perreault, 1999).

Lastly, climate change has inspired economic geographers to examine how communities and economies will adapt to global warming and how greenhouse gases might be more effectively regulated. Although climate change is a global phenomenon, its effects will vary significantly from region to region and will require a diverse range of socioeconomic adaptation strategies (Yohe and Schlesinger, 2002). Poor communities and regions will be particularly vulnerable as they experience what Leichenko and O'Brien (2008) term 'double exposure', a situation where the security of people's livelihoods are simultaneously threatened by the challenges of economic globalization and the environmental impacts of climate change. Effective collective action at the local and global scales is crucial if there is to be a fair and progressive distribution of the costs of, and responsibilities for, greenhouse gas regulation and climate change adaptation (Adger, 2003; O'Brien and Leichenko, 2006; Bumpus and Liverman, 2008).

222

KEY POINTS

- The concept of sustainable development integrates ecological, social and economic concerns into a single developmental model that strives to ensure inter-generational equity with respect to material, social and environmental well-being.
- Weak proponents of sustainable development argue that a growth-oriented economic system is compatible with sustainability while strong proponents call for a radical transformation to a steady-state economic system.
- There are significant critiques of sustainable development, particularly from political ecology studies, that highlight the concept's ambiguity, the power structures that promote it and the negative consequences of sustainability policies and programmes.
- Environmental economic geographers study the sustainability of industries, the sustainability of urban and rural livelihoods, and the links between global climate change, economic globalization and socio-economic vulnerability.

FURTHER READING

Soyez and Schultz (2008) introduce a special issue (with five separate papers) in the journal *Geoforum* that focuses on the subfield of environmental economic geography. Krueger and Gibbs's (2007) volume provides a critical assessment of the sustainable development concept as it relates to urban planning issues in Europe and the USA. See Bunce (2008) for a case study examining the sustainability of Barbados's tourism-based economy and for a discussion of the challenges facing small-island economies.

NOTES

1 The Rev. Thomas Malthus (1766–1834) was a British philosopher who argued in his book *The Principle of Population* that population growth would inevitably outpace the earth's ability to provide for greater numbers of people unless 'positive' factors (e.g. higher death rates caused by food scarcity) and 'preventative' checks (e.g. birth control) maintained more sustainable levels of population.

2 Related studies have examined if and how global value chains are becoming 'greener' through new regulatory and certification systems (e.g. see Ponte, 2008; 4.4 Global Value Chains) and assessed the sustainability of resource peripheries (i.e. rural places rich in natural resources) that provide raw materials for the global economy (e.g. see Hayter et al., 2003; 4.1 Core–Periphery).

223

REFERENCES

Introduction

Aglietta, M. (1979) *A Theory of Capitalist Regulation: The US Experience*. An English translation of *Régulation et Crises du Capitalisme*. London: New Left Books (originally published in 1976).

Alonso, W. (1964) *Location and Land Use*. Cambridge, MA: Harvard University Press.

Atwood, W. W. (1925) 'Economic geography.' *Economic Geography* 1: 1 (March).

Barnes, T. J (2000) 'Inventing Anglo-American Economic Geography, 1889–1960.' Chapter in E. Sheppard and T. Barnes (eds), *A Companion to Economic Geography*. New York: Blackwell, pp.11–26.

Barnes, T., Sheppard, E., Tickell, A. and J. Peck (eds) (2007) *Politics and Practice in Economic Geography*. London: Sage Publications.

Berry, B.J.L. & Garrison, W.L. (1958) 'The functional bases of the central place hierarchy,' *Economic Geography* 34(2): 145–54.

Bluestone, B. and Harrison, B. (1982) *The Deindustrialization of America: Plant Closings, Community Abandonment, and the Dismantling of Basic Industry*. New York: Basic Books.

Castells, M. (1984) *City and the Grassroots: A Cross-cultural Theory of Urban Social Movements*. Berkeley and Los Angeles, CA: University of California Press.

Castells, M. (ed) (1985) *High Technology, Economic Restructuring, and the Urban-Regional Process in the United States*. Thousand Oaks, CA: Sage Publications.

Chandler, A. D., Jr. (1977) *The Visible Hand: The Managerial Revolution in American Business:* Harvard University Press.

Chojnicki, Z. (1970) 'Prediction in economic geography,' *Economic Geography*, 46S (Proceedings, International Geographic Union, Commission on Quantitative Methods): 213–22.

Clark, G.L. (1986) 'Restructuring the U.S. Economy: The NLRB, the Saturn Project and economic justice', *Economic Geography* 62(4): 289–306.

Combes, P.-P., Mayer, T. and Thisse, J.-F. (2008) *Economic Geography: The Integration of Regions and Nations*. Princeton, NJ: Princeton University Press.

Dear, M. (1988) 'The postmodern challenge: Reconstructing human geography', *Transactions, Institute of British Geographers* 13(3): 262–74.

Fisher, C.A. (1948) 'Economic geography for a changing world,' *Transactions of the Institute of British Geographers* 14: 71–85.

Frank, A.G. (1966) 'The development of underdevelopment', *Monthly Review* 18: 17–31.

Gereffi, G. (1994) 'The organization of buyer-driven global commodity chains: How U.S. retailers shape overseas production networks', in G. Gereffi and M. Koreniewicz (eds) *Commodity Chains and Global Capitalism*. Westport, CT: Praeger. pp. 95–122.

References

Giddens, A. (1984) *The Constitution of Society*. Cambridge: Polity Press.

Granovetter, M.S. (1985) 'Economic action and social structure: The problem of embeddedness', *American Journal of Sociology* 91(3): 481–510.

Hartshorne, R. (1939) *The Nature of Geography: A Critical Survey of Current Thought in the Light of the Past*. Philadelphia: Association of American Geographers.

Harvey, D. (1968) 'Some methodological problems in the use of the Neyman Type A and the negative binomial probability distributions for the analysis of spatial point patterns,' *Transactions of the Institute of British Geographers* 44: 85–95.

Harvey, D. (1969) *Explanation in Geography*. London: Edward Arnold. pp. 113–129.

Harvey, D. (1974) 'What kind of geography for what kind of public policy?' *Transactions of the Institute of British Geographers* 63: 18–24.

Harvey, D. (1989) *The Condition of Postmodernity: An Enquiry into the Origins of Cultural Change*. Oxford: Basil Blackwell.

Hoover, E.M. (1948) *The Location of Economic Activity*. New York: McGraw-Hill. pp. 1–185.

Huntington, E. (1940) *Principles of Economic Geography*. New York: Wiley & Sons.

Isard, W. (1949) 'The general theory of location and space-economy,' *The Quarterly Journal of Economics* 63(4): 476–506.

Isard, W. (1953) 'Regional commodity balances and interregional commodity flows,' *The American Economic Review* 43(2): 167–180.

Keasbey, L.M. (1901a) 'The study of economic geography,' *Political Science Quarterly* 16(1): 79–95.

Keasbey, L.M. (1901b) 'The principles of economic geography,' *Political Science Quarterly* 16(3): 472–481.

Krugman, P. (1991) 'Increasing returns and economic geography', *Journal of Political Economy* 99(3): 483–499.

Malecki, E. J. (1985) 'Industrial location and corporate organization in high technology industries', *Economic Geography* 61(4): 345–369.

Nelson, R.R. and Winter, S.G. (1974) 'Neoclassical vs. evolutionary theories of economic growth: Critique and prospectus', *The Economic Journal* (December): 886–905.

Piore, M.J. and Sabel, C.F. (1984) *The Second Industrial Divide: Possibilities for Prosperity*. New York: Basic Books.

Schoenberger, E. (1985) 'Foreign manufacturing investment in the United States: Competitive strategies and international location', *Economic Geography* 61(3): 241–259.

Scott, A.J. (1969) 'A model of spatial decision-making and locational equilibrium,' *Transactions of the Institute of British Geographers* 47: 99–110.

Scott, A. J. and M. Storper (eds) (1986) *Production, Work, Territory: The Geographical Anatomy of Industrial Capitalism*. Boston: Allen & Unwin.

Smith, J.R. (1907) 'Economic geography and its relation to economic theory and higher education,' *Bulletin of the American Geographical Society* 39(8): 472–481.

Smith, J.R. (1913) *Industrial and Commercial Geography*. Henry Holt & Company: New York.

Soja, E. (1989) *Postmodern Geographies: The Reassertion of Space in Critical Social Theory*. London: Verso.

Wallerstein, I. (1974) *The Modern World-System: Capitalist Agriculture and the Origins of the European World-Economy in the Sixteenth Century.* New York: Academic Press.

Williamson, O. (1981) 'The modern corporation: Origins, evolution, attributes,' *Journal of Economic Literature* 19 (December): 1537–1568.

1 Key Agents in Economic Geography

1.1 Labour

Boschma, R., Eriksson, R. and Lindgren, U. (2009) 'How does labour mobility affect the performance of plants? The importance of relatedness and geographical proximity', *Journal of Economic Geography* 9(2): 169–190.

Cho, S.K. (1985) 'The labour process and capital mobility: the limits of the new international division of labour', *Politics and Society* 14: 185–222.

Christopherson, S. (1983) 'The household and class formation: determinants of residential location in Ciudad Juarez', *Environment and Planning D: Society and Space* 1: 323–338.

Christopherson, S., and Lillie, N. (2005) 'Neither global nor standard: corporate strategies in the new era of labor standards', *Environment and Planning A* 37: 1919–1938.

Dicken, P. (1971) 'Some aspects of decision-making behavior of business organizations', *Economic Geography* 47: 426–437.

Florida, R. (2002). *The Rise of the Creative Class: And How It's Transforming Work, Leisure, Community and Everyday Life.* New York: Basic Books.

Glassman, J. (2004) 'Transnational hegemony and US labor foreign policy: towards a Gramscian international labour geography', *Environment and Planning D: Society and Space* 22: 573–593.

Hamilton, T. (2009) 'Power in numbers: a call for analytical generosity toward new political strategies', *Environment and Planning A* 41: 284–301.

Harvey, D. (1982) *The Limits to Capital.* London: Verso.

Herod, A. (1997) 'From a geography of labor to a labor geography: labor's spatial fix and the geography of capitalism', *Antipode* 29(1): 1–31.

Herod, A. (2002) *Labour Geographies: Workers and the Landscapes of Capitalism.* NewYork: Guilford.

Holmes, J. (2004) 'Re-scaling collective bargaining: union responses to restructuring in the North American auto industry', *Geoforum* 35(1): 9–21.

Hughes, A., Wrigley, N. and Buttle, M. (2008) 'Global production networks, ethical campaigning, and the embeddedness of responsible governance', *Journal of Economic Geography* 8(3): 345–367.

Kuemmerle, W. (1999) 'Foreign direct investment in industrial research in the pharmaceutical and electronics industries – results from a survey of multinational firms', *Research Policy* 28(2–3): 179–193.

References

Malecki, E.J. and Moriset, B. (2008) *The Digital Economy: Business Organization, Production Processes, and Regional Developments*. Abingdon, UK and New York: Routledge.

Marx, K. (1867) *Das Capital: Kritik der politischen Oekonomie. Buch 1: Der Produktionsprocess des Kapitals*. Hamburg: Verlag von Otto Meissner.

Massey, D. (1984) *Spatial Divisions of Labour: Social Structures and the Geography of Production*. New York: Methuen.

McDowell, L. (1991) 'Life without father and Ford', *Transactions of the Institute of British Geographers* 16: 400–419.

Nakano-Glenn, E. (1985) 'Racial ethnic women's labour: the intersection of race, gender, and class oppression', *Review of Radical Political Economy* 17: 86–106.

Peck, J. (1996) *Work-Place: The Social Regulation of Labour Markets*. New York: Guilford Press.

Peck, J. (2001) *Workfare States*. New York: Guilford Press.

Peck, J. and Theodore, N. (2001) 'Contingent Chicago: restructuring the spaces of temporary labour', *International Journal of Urban and Regional Research* 25(3): 471–496.

Peet, J.R. (1975) 'Inequality and poverty: a Marxist-geographic theory', *Annals of the Association of American Geographers* 65(4): 564–571.

Ricardo, D. (1817) *Principles of Political Economy and Taxation*. London: John Murray.

Riisgaard, L. (2009) 'Global value chains, labour organization and private social standards: lessons from East African cut flower industries', *World Development* 37(2): 326–340.

Rutherford, T.D. and Gertler, M.S. (2002) 'Labour in "lean" times: geography, scale and the national trajectories of workplace change', *Transactions of the Institute of British Geographers* 27(2): 195–212.

Savage, L. (2006) 'Justice for janitors: scales of organizing and representing workers', *Antipode* 38(3): 648–667.

Savage, L. and Wills, J. (2004) 'New geographies of trade unionism', *Geoforum* 35(1): 5–7.

Saxenian, A. (2006) *The New Argonauts: Regional Advantage in a Global Economy*. Cambridge, MA: Harvard University Press.

Smith, A. (1776). *An Inquiry into the Nature and Causes of the Wealth of Nations*. London: W. Strahan and T. Cadell.

Storper, M. and Walker, R. (1983) 'The theory of labour and the theory of location', *International Journal of Urban and Regional Research* 7: 1–43.

Vind, I. (2008) 'Transnational companies as a source of skill upgrading: the electronics industry in Ho Chi Minh City', *Geoforum* 39(3): 1480–1493.

Walsh, J. (2000) 'Organizing the scale of labour regulation in the United States: service sector activism in the city', *Environment and Planning A* 32: 1593–1610.

Walker, R. and Storper, M. (1981) 'Capital and industrial location', *Progress in Human Geography* 7(1): 1–41.

Weber, A. (1929) *Theory of the Location of Industries*. An English translation of *Über den Standort der Industrien* by C.J. Friedrich. Chicago: University of Chicago Press (originally published in 1909).

Wills, J. (1998) 'Taking on the CosmoCorps? Experiments in transnational labour organization', *Economic Geography* 74(2) (April): 111–130.

1.2 Firm

Angel, D.P. and Savage, L.A. (1996) 'Global localization? Japanese research and development laboratories in the USA', *Environment and Planning A* 28(5): 819–833.

Aoyama, Y. (2000) 'Keiretsu, networks and locations of Japanese electronics industry in Asia', *Environment and Planning A* 2 (February): 223–244.

Barney, J. (1991) 'Firm resources and sustained competitive advantage', *Journal of Management* 17(1): 99–120.

Barney, J., M. Wright, and Ketchen, J. (2001) 'The resource-based view of the firm: ten years after 1991', *Journal of Management* 27(6): 625–641.

Beugelsdijk, S. (2007) 'The regional environment and a firm's innovative performance: a plea for a multilevel interactionist approach', *Economic Geography* 83(2): 181–199.

Chandler, A.D., Jr. (1962) *Strategy and Structure: Chapters in the History of the American Industrial Enterprise.* Cambridge, MA: MIT Press.

Chandler, A.D., Jr. (1977) *The Visible Hand: The Managerial Revolution in American Business.* Harvard University Press.

Coase, R.H. (1937) 'The nature of the firm', *Economica* 4(16): 386–405.

Denzau, A.T. and North, D.C. (1994) 'Shared mental models: ideologies and institutions', *Kyklos* 47(1): 3–31.

Dicken, P. (1971) 'Some aspects of decision-making behaviour in business organizations', *Economic Geography* 47: 426–437.

Dicken, P. (1976) 'The multiplant business enterprise and geographical space', *Regional Studies* 10: 401–412.

Dicken, P. and Lloyd, P. (1980) 'Patterns and processes of change in the spatial distribution of foreign-controlled manufacturing employment in the United Kingdom, 1963–1975', *Environment and Planning A* 12: 1405–1426.

Dicken, P. and Malmberg, A. (2001) 'Firms in territories: a relational perspective', *Economic Geography* 77(4): 345–363.

Donaghu, M.T and Barff, R. (1990) 'Nike just did it: international subcontracting and flexibility in athletic footwear production', *Regional Studies* 24(6): 537–552.

Dunning, J.H. (1977) 'Trade, location of economic activity and the multinational enterprise: A search for an eclectic approach', in B. Ohlin, P.O. Hesselborn and P.M. Wijkman (eds), *The International Allocation of Economic Activity.* London: Macmillan, pp. 395–418.

Encarnation, D.J. and Mason, M. (eds) (1994) *Does Ownership Matter?: Japanese Multinationals in Europe.* Oxford: Oxford University Press.

Fields, G. (2006) 'Innovation, time and territory: space and the business organization of Dell computer', *Economic Geography* 86(2): 119–146.

Florida, R. and Kenney, M. (1994) 'The globalization of Japanese R&D: the economic geography of Japanese R&D investment in the United States', *Economic Geography* 70(4): 344–369.

Grabher, G. (2002) 'The project ecology of advertising: tasks, talents and teams', *Regional Studies* 36(3): 245–262.

Harrison, B. (1992) 'Industrial districts: old wine in new bottles', *Regional Studies* 26(5): 469–483.

References

Holloway, S.R. and Wheeler, J.O. (1991) 'Corporate headquarters relocation and changes in metropolitan corporate dominance, 1980–1987', *Economic Geography* 67(1): 54–74.

Hughes, A., Buttle, M. and Wrigley, N. (2007) 'Organisational geographies of corporate responsibility: a UK–US comparison of retailers' ethical trading initiatives', *Journal of Economic Geography* 7(4): 491–513.

Lee, Y.-S. (2003) 'Lean production systems, labour unions, and greenfield locations of Korean new assembly plants and their suppliers', *Economic Geography* 79(3): 321–339.

Markusen, A.R. (1985) *Profit Cycles, Oligopoly and Regional Development.* Cambridge, MA: MIT Press.

Maskell, P. (1999) 'The firm in economic geography', *Economic Geography* 77(4) (October): 329–344.

Maskell, P. and Malmberg, A. (2001) 'The competitiveness of firms and regions: "ubiquitification" and the importance of localized learning', *European Urban and Regional Studies* 6(1): 9–25.

Monk, A. (2008) 'The knot of contracts: the corporate geography of legacy costs', *Economic Geography* 84(2): 211–235.

Penrose, E.T. (1959) *The Theory of the Growth of the Firm.* Oxford: Oxford University Press.

Piore, M.J. and Sabel, C.F. (1984) *The Second Industrial Divide: Possibilities for Prosperity.* New York: Basic Books.

Phelps, N.A. (2000) 'The locally embedded multinational and institutional capture', *Area* 32(2): 169–178.

Prahalad, C.K. (1993) 'The role of core competencies in the corporation', *Research Technology Management* 36(6) (November/December): 40–47.

Reich, R.B. (1990) 'Who Is Us?' *Harvard Business Review* (January–February): 53–64.

Riordan, M.H. and Williamson, O.E. (1985) 'Asset specificity and economic organization', *International Journal of Industrial Organization* 3: 365–378.

Sabel, CF (1993) 'Studied trust: building new forms of cooperation in a volatile economy', *Human Relations* 46(9): 1133–1170.

Sako, M. (1992) *Prices, Quality and Trust: Inter-firm Relations in Britain and Japan.* Cambridge: Cambridge University Press.

Schoenberger, E. (1997) *The Cultural Crisis of the Firm.* Cambridge, MA: Blackwell Publishers.

Scott, A.J. (1988) *New Industrial Spaces: Flexible Production Organization and Regional Development in North America and Western Europe.* London: Pion.

Simon, H.A. (1947) *Administrative Behaviour.* New York: Macmillan.

Törnqvist, G. (1968) Flows of Information and the Location of Economic Activities. *Geografiska Annaler. Series B, Human Geography* 50(1): 99–107.

Tyson, L.D. (1991) 'They are not us: why American ownership still matters', *American Prospect* 4(Winter): 37–49.

Watts, H. (1980) *The Large Industrial Enterprise.* London: Croom Helm.

Williamson, O.E. (1981) 'The modern corporation: origins, evolution, attributes', *Journal of Economic Literature* XIX(December): 1537–1569.

229

Williamson, O.E. (1993) 'Calculativeness, trust, and economic organization', *Journal of Law and Economics* 36(1): 453–486.

Yeung, H.W.-c. (1997) 'Business networks and transnational corporations: a study of Hong Kong firms in the ASEAN region', *Economic Geography* 73(1): 1–25.

Yeung, H.W.-c. (1999) 'The internationalization of ethnic Chinese business firms from Southeast Asia: strategies, processes and competitive advantage', *International Journal of Urban and Regional Research* 23(1): 103–127.

Yeung, H.W.-c. (2000) 'Organising "the firm" in industrial geography I: networks, institutions and regional development', *Progress in Human Geography* 24(2): 301–315.

1.3 State

Albert, M. (1993) *Capitalism Against Capitalism*. London: Whurr.

Amsden, A.H. (1989) *Asia's Next Giant: South Korea and Late Industrialization*. New York: Oxford University Press.

Arndt, H.W. (1988) '"Market failure" and underdevelopment', *World Development* 16(2): 219–229.

Arrighi, G. (2002) 'The African crisis: World systemic and regional aspects', *New Left Review* 15: 5–36.

Berger, S. and Dore, R. (eds) (1996) *National Diversity and Global Capitalism*. Ithaca, NY: Cornell University Press.

Brenner, N. (2004) *New State Spaces: Urban Governance and the Rescaling of Statehood*. Oxford: Oxford University Press.

Brohman, J. (1996) *Popular Development: Rethinking the Theory and Practice of Development*. Oxford: Blackwell.

Clark, G.L. and Dear, M.J. (1984) *State Apparatus: Structures of Language and Legitimacy*. Boston: Allen and Unwin.

Cox, K. (2002) *Political Geography: Territory, State, and Society*. Oxford: Blackwell.

Das, R.J. (1998) 'The social and spatial character of the Indian State', *Political Geography* 17(7): 787–808.

Dicken, P. (2007) *Global Shift: Mapping the Changing Contours of the World Economy*. 5th edn. New York: Guilford Press.

Evans, P.B. (1995) *Embedded Autonomy: States and Industrial Transformation*. Princeton, NJ: Princeton University Press.

Gilbert, A. (2007) 'Inequality and why it matters', *Geography Compass* 1(3): 422–447.

Glassman, J. (1999) 'State power beyond the "territorial trap": the internationalization of the state', *Political Geography* 18(6): 669–696.

Glassman, J. and Samatar. A.I. (1997) 'Development geography and the Third World state', *Progress in Human Geography* 21(2): 164–198.

Hall, P.A. and D. Soskice (eds) (2001) *Varieties of Capitalism: The Institutional Foundations of Comparative Advantage*. Oxford: Oxford University Press.

Harvey, D. (2005) *A Brief History of Neoliberalism*. New York: Oxford University Press.

Haughwout, A.F. (1999) 'State infrastructure and the geography of employment', *Growth and Change* 30(4): 549–566.

Hindery, D. (2004) 'Social and environmental impacts of World Bank/IMF-funded economic restructuring in Bolivia: an analysis of Enron and Shell's hydrocarbons projects', *Singapore Journal of Tropical Geography* 25(3): 281–303.

Hollander, G. (2005) 'Securing sugar: national security discourse and the establishment of Florida's sugar-producing region', *Economic Geography* 81(4): 339–358.

Jessop, B. (1990) *State Theory: Putting Capitalist States in their Place.* University Park, PA: Pennsylvania State University Press.

Jessop, B. (1994) 'Post-Fordism and the state', in A. Amin (ed.) *Post-Fordism: A Reader,* Oxford: Blackwell. pp. 251–279.

Johnson, C. (1982) *MITI and the Japanese Miracle: The Growth of Industrial Policy 1925–1975.* Palo Alto, CA: Stanford University Press.

Jones, M. (2001) 'The rise of the regional state in economic governance: "partnerships for prosperity" or new scales of state power?', *Environment and Planning A* 33(7): 1185–1211.

Lall, S. (1994) 'The East Asian miracle: does the bell toll for industrial strategy?', *World Development* 22(4): 645–654.

Lonsdale, R.E. (1965) 'The Soviet concept of the territorial-production complex', *Slavic Review* 24(3): 466–478.

Manger, M. (2008) 'International investment agreements and services markets: Locking in market failure?', *World Development* 36(11): 2456–2469.

O'Neill, P. (1997) 'Bringing the qualitative state into economic geography', in R. Lee and J. Wills (eds) *Geographies of Economies.* London: Arnold. pp.290–301.

Peck, J. (2001) *Workfare States.* New York: Guilford Press.

Peck, J. and Theodore, N. (2007) 'Variegated capitalism', *Progress in Human Geography* 31(6): 731–772.

UNCTAD (United Nations Conference on Trade and Development) (2005) *World Investment Report 2008: Transnational Corporations and the Internationalization of R&D.* Geneva: UNCTAD.

World Bank (1994) *The East Asian Miracle: Economic Growth and Public Policy.* Washington, DC: World Bank.

2 Key Drivers of Economic Change

2.1 Innovation

Arrow, K. (1962) 'The economic implications of learning-by-doing', *Review of Economic Studies* 29(3): 155–173.

Arthur, B.W. (1989) 'Competing technologies, increasing returns and lock-in by historical events', *The Economic Journal* 99(394): 116–131.

Blake, M. and Hanson, S. (2005) 'Rethinking innovation: context and gender', *Environment and Planning A* 37: 681–701.

Boschma, R. and Frenken, K. (2009) 'Some notes on institutions in evolutionary economic geography', *Economic Geography* 85(2): 151–158.

Boschma, R.A. and Lambooy, J.G. (1999) 'Evolutionary economics and economic geography', *Journal of Evolutionary Economics* 9: 411–429.

Boschma, R.A. and Martin, R. (2007), 'Constructing an evolutionary economic geography', *Journal of Economic Geography* 7(5): 537–548.

Brown, L. (1981) *Innovation Diffusion: A New Perspective.* London: Methuen.

References

David, P.A. (1985) 'Clio and the Economics of QWERTY', *American Economic Review* 75(2): 332–337.

Domar, E. (1946) 'Capital expansion, rate of growth, and employment', *Econometrica* 14(April): 137–147.

Dosi, G. (1982) 'Technological paradigms and technological trajectories: a suggested interpretation of the determinants and directions of technical change', *Research Policy* 11(3): 147–162.

Dosi, G. (1997) 'Opportunities, incentives and the collective patterns of technical change', *Economic Journal* 107(444) (September): 1530–1547.

Essletzbichler, J.(2009) 'Evolutionary economic geography, institutions, and political economy', *Economic Geography* 85(2):159–165.

Essletzbichler, J. and Rigby, D.L. (2007) 'Exploring evolutionary economic geographies', *Journal of Economic Geography* 7(5): 549–571.

Feldman, M.P. and Massard, N. (eds) (2001) *Institutions and Systems in the Geography of Innovation*. Boston: Kluwer Academic Publishers.

Freeman, C. (1974) *The Economics of Industrial Innovation*. Harmondsworth: Penguin.

Freeman, C. (1988) *Technical Change and Economic Theory*. London: Pinter.

Freeman, C. (1991) 'Innovation, changes of techno-economic paradigm and biological analogies in economics', *Revue Economique* 42(2): 211–231.

Freeman, C. (1995) 'The "National System of Innovation" in historical perspective', *Cambridge Journal of Economics* 19: 5–24.

Gertler, M. (2003) 'Tacit knowledge and the economic geography of context, or the undefinable tacitness of being (there)', *Journal of Economic Geography* 3(1): 75–99.

Grabher, G. (2009) 'Yet another turn? The evolutionary project in economic geography, *Economic Geography* 85(2):119–127.

Grabher, G., Ibert, O. and Floher, S. (2008) 'The neglected king: the customer in the new knowledge ecology of innovation', *Economic Geography* 84(3): 253–280.

Hall, P. and Preston, P. (1988) 'The long-wave debate', Chapter 2 in *The Carrier Wave*. London: Unwin Hyman: 12–27.

Hägerstrand, T. (1967) *Innovation Diffusion as a Spatial Process*. Chicago: University of Chicago Press.

Harrod, R.F. (1948) *Toward a Dynamic Economcis: Some Recent Developments of Economic Theory and Their Applications to Policy*. London: Macmillan.

Hodgson, G.M. (2009) 'Agency, institutions, and Darwinism in evolutionary economic geography', *Economic Geography* 85(2):167–173.

Kondratieff, N. (1926) Die langen Wellen der konjuktur. *Archiv für Socialwissenschaft* 56: 573–609.

Kondratieff, N. (1935) 'The long waves in economic life', *The Review of Economic Statistics* 17: 105–115.

Kuznets, S.S. (1940) 'Schumpeter's business cycles', *American Economic Review* 30: 250–271.

Lundvall, B.-Å. (1988) 'Innovation as an interactive process: from user–producer interaction to the national system of innovation', in G. Dosi, C. Freeman, R. Nelson, G. Silverberg and L. Soete (eds) *Technical Change and Economic Theory*. London: Pinter pp. 349–369.

Lundvall, B-Å. and Johnson, B. (1994) 'The learning economy', *Industry and Innovation* 1(2): 23–42.

MacKinnon, D., Cumbers, A., Pike, A., Birch, K. and R. McMaster (2009) 'Evolution in economic geography: institutions, political economy, and adaptation', *Economic Geography* 85(2): 129–150.

Martin, R. (2010) 'Rethinking regional path-dependence: beyond lock-in to evolution', *Economic Geography* 86(1): 1–27.

Martin, R. and Sunley, P. (2006) 'Path dependence and regional economic evolution', *Journal of Economic Geography* 6(4): 395–437.

Nelson, R.R. and Winter, S.G. (1974) 'Neoclassical vs. evolutionary theories of economic growth: critique and prospectus', *The Economic Journal* 84(336): 886–905.

Pike, A., Birch, K., Cumbers, A., MacKinnon, D. and McMaster, R. (2009) 'A geographical political economy of evolution in economic geography', *Economic Geography* 85(2): 175–182.

Polanyi, M. (1967) *The Tacit Dimension*. Chicago: University of Chicago Press.

Porter, M. (1990) *The Competitive Advantage of Nations*. New York: Free Press.

Rogers, E,M. (1962) *Diffusion of Innovations*. New York: The Free Press.

Rosenberg, N. (1982) *Inside the Black Box: Technology and Economics*. Cambridge: Cambridge University Press.

Schumpeter, J. (1928) 'The instability of capitalism', *The Economic Journal* 38(151): 361–386.

Schumpeter, J. (1939) *Business Cycles: A Theoretical, Historical and Statistical Analysis of the Capitalist Process*, 2 vols. New York: McGraw-Hill.

Schumpeter, J.A. (1942) *Capitalism, Socialism and Democracy*. New York: Harper.

Solow, R. (1957) 'Technical change and the aggregate production function', *Review of Economics and Statistics* 39: 312–320.

Storper, M. (1997) *The Regional World: Territorial Development in a Global Economy*. New York: Guilford.

Utterback, J. and Abernathy, W. (1975) 'A dynamic model of process and product innovation', *Omega* 3(6): 639–656.

von Hippel, E. (1976) 'The dominant role of users in the scientific instrument innovation process', *Research Policy* 5(3): 212–239.

von Hippel, E. (2005) *Democratizing Innovation*. Cambridge, MA: MIT Press.

233

2.2 Entrepreneurship

Acs, Z. and Audretsch, A. (2003) 'Introduction', in Z. Acs and D. Audretsch (eds) *Handbook of Entrepreneurship Research*. Boston: Kluwer Academic Publishers. pp. 3–20.

Acs, Z., Carlsson, B. and Karlsson, C. (eds) (1999) *Entrepreneurship, Small and Medium-Sized Enterprises and the Macroeconomy*. Cambridge: Cambridge University Press.

Aoyama, Y. (2009) 'Entrepreneurship and regional culture: the case of Hamamatsu and Kyoto, Japan', *Regional Studies* 43(3): 495–512.

Bengtsson, M. and Soderholm, A. (2002) 'Bridging distances: organizing boundary-spanning technology development projects', *Regional Studies* 36(3): 263–274.

Birch, D. (1981) 'Who creates jobs?', *The Public Interest* 65: 3–14.

Birley, S. (1985) 'The role of networks in the entrepreneurial process', *Journal of Business Venturing* 1: 107–117.

Blake, M. (2006) 'Gendered lending: gender, context and the rules of business lending', *Venture Capital* 8(2): 183–201.

References

Blake, M. and Hanson, S. (2005) 'Rethinking innovation: context and gender', *Environment and Planning A* 37: 781–701.

Bosma, N.S. and Schutjens, V. (2007) 'Outlook on Europe: patterns of promising entrepreneurial activity in European regions', *Tijdschrift voor economische en sociale geografie* 98(5): 675–686.

Bosma, N.S. and Schutjens, V. (2009) 'Mapping entrepreneurial activity and entrepreneurial attitudes in European regions', *International Journal of Entrepreneurship and Small Business* 9(2).

Bosma, N.S., Acs, Z.J., Autio, E., Coduras, A. and Le, J. (2008) *Global Entrepreneurship Monitor: 2008 Executive Report.* Babson College, Universidad del Desarrollo (Santiago, Chile) and London Business School (London).

Chinitz, B. (1961) *Economic Study of the Pittsburgh Region.* Pittsburgh Regional Plan Association, Pittsburgh, Pennsylvania.

de Soto, H. (1989) *The Other Path: The Invisible Revolution in the Third World.* New York: HarperCollins.

Flora, J., Sharp, J., Flora, C. and Newlon, B. (1997) 'Entrepreneurial social infrastructure and locally initiated economic development in the nonmetropolitan United States', *Sociological Quarterly* 38(4): 623–645.

Gartner, W. and Shane, S. (1995) 'Measuring entrepreneurship over time', *Journal of Business Venturing* 10: 283–301.

Greenbaum, R. and Tita, G. (2004) 'The impact of violence surges on neighborhood business activity', *Urban Studies* 4(13): 2495–2514.

Hanson, S. and Blake, M. (2009) 'Gender and entrepreneurial networks', *Regional Studies* 43(1): 135–149.

Hess, M. (2004) '"Spatial" relationships? Towards a reconceptualization of embeddedness', *Progress in Human Geography* 28(2): 165–186

Jack, S. and Anderson, A. (2002) 'The effects of embeddedness on the entrepreneurial process', *Journal of Small Business Venturing* 17(5): 467–488.

Kalantaridis, C. and Zografia, B. (2006) 'Local embeddedness and rural entrepreneurship: case study evidence from Cumbria, England', *Environment and Planning A* 38: 1561–1579.

Keeble, D. and Walker, S. (1994) 'New firms, small firms, and dead firms: spatial patterns and determinants in the UK', *Regional Studies* 28(4): 411–427.

Kenney, M. and Patton, D. (2005) 'Entrepreneurial geographies: support networks in three high-technology industries', *Economic Geography* 8(2): 201–228.

Kilkenny, M., Nalbarte, L. and Besser, T. (1999) 'Reciprocated community support and small town-small business success', *Entrepreneurship and Regional Development* 11: 231–246.

Kirzner, I.M. (1973) *Competition and Entrepreneurship.* Chicago: University of Chicago Press.

Malecki, E. (1997) *Technology and Economic Development: The Dynamics of Local, Regional, and National Change.* New York: John Wiley.

Malecki, E.J. (1994) 'Entrepreneurship in regional and local development', *International Regional Science Review* 16(1 & 2): 119–153.

Malecki, E.J. (1997) 'Entrepreneurs, networks, and economic development: a review of recent research', in J.A. Katz and R.H. Brockhaus (eds) *Advances in Entrepreneurship, Firm Emergence and Growth* (Vol. 3). Greenwich, CT: JAI Press Inc. pp. 57–118.

References

Mandel, J. (2004) 'Mobility matters: women's livelihood strategies in Porto Novo, Benin', *Gender, Place, and Culture* 1(2): 257–287.

Murphy, J.T. (2006) 'The socio-spatial dynamics of creativity and production in Tanzanian industry: urban furniture manufacturers in a liberalizing economy', *Environment and Planning A* 38(10): 1863–1882.

Nijkamp, P. (2003) 'Entrepreneurship in a modern network economy', *Regional Studies* 37: 395–405.

Pallares-Barbera, M., Tulla, A. and Vera, A. (2004) 'Spatial loyalty and territorial embeddedness in the multi-sector clustering of the Bergued, a region in Catalonia (Spain)', *Geoforum* 35: 635–649.

Reynolds, P. (1991) 'Sociology and entrepreneurship: concepts and contributions', *Entrepreneurship Theory and Practice* 15: 47–70.

Reynolds, P. and White, S.B. (1997) *The Entrepreneurial Process.* New London, CT: Quorum Books.

Sarasvathy, S., Dew, N., Ramakrishna, S. and Venkataraman, S. (2003) 'Three views of entrepreneurial opportunity', in Z. Acs and D. Audretsch (eds) *Handbook of Entrepreneurship Research: An Interdisciplinary Survey and Introduction.* Boston: Kluwer Academic Publishers. pp. 141–160.

Saxenian, A. (1994) *Regional Advantage: Culture and Competition in Silicon Valley and Route 128.* Cambridge, MA: Harvard University Press.

Saxenian, A. (2006) *The New Argonauts: Regional Advantage in a Global Economy.* Cambridge, MA: Harvard University Press.

Schoonhoven, C.B. and Romanelli, E., eds. (2001) *The Entrepreneurship Dynamic: Origins of Entrepreneurship and the Evolution of Industries.* Stanford, CA: Stanford Business Books.

Schumpeter, J.A. (1942) *Capitalism, Socialism, and Democracy.* New York: Harper & Brothers.

Schumpeter, J.P. (1936) *The Theory of Economic Development: An Inquiry into Profits, Capital, Credit, Interest, and the Business Cycle.* Trans. from the German by R. Opie. Cambridge, MA: Harvard University Press.

Schutjens, V. and Stam, E. (2003) 'The evolution and nature of young firm networks: a longitudinal perspective', *Small Business Economics* 21: 115–134.

Shane, S. and Eckhardt, J. (2003) 'The individual–opportunity nexus', in Z.J. Acs and D.B. Audretsch (eds) *Handbook of Entrepreneurship Research.* Norwell, MA: Kluwer Academic Publishers. pp. 161–191.

Shaw, E. (1997) 'The "real" networks of small firms', In D. Deakns, P. Jennings and C. Mason (eds), *Small Firms: Entrepreneurship in the Nineties.* London: Paul Chapman Publishing Ltd. pp. 7–17.

Sorenson, O. and Baum, J. (2003) 'Editors' introduction: geography and strategy – the strategic management of space and place', *Advances in Strategic Management* 20: 1–19.

Stam, E. (2007) 'Why butterflies don't leave: locational behavior of entrepreneurial firms', *Economic Geography* 83(1): 27–50.

Stevenson, H. (1999) 'A perspective on entrepreneurship', in W.A. Sahlman, H.H. Stevenson, and M.J. Roberts (eds) *The Entrepreneurial Venture.* Boston, MA: Harvard Business School Press.

Thornton, P.H. (1999) 'The sociology of entrepreneurship', *Annual Review of Sociology* 25: 19–46.

Thornton, P. and Flynn, K. (2003) 'Entrepreneurship, networks, and geographies', in Z. Acs and D. Audretsch (eds) *Handbook of Entrepreneurship Research: An Interdisciplinary Survey and Introduction*. Boston: Kluwer Academic Publishers. pp. 401–433.

Turner, S. (2007) 'Small-scale enterprise livelihoods and social capital in Eastern Indonesia: ethnic embeddedness and exclusion', *The Professional Geographer* 59(4): 407–420.

United States Census Bureau (2009) 'Statistics about business size (including Small Business) from the U.S. Census Bureau': http://www.census.gov/epcd/www/small-bus.html (accessed on 17 September 2009).

Waldinger, R., Aldrich, H. and Ward, R. (1990) *Ethnic Entrepreneurs: Immigrant Businesses in Industrial Societies*. Newbury Park, CA: Sage Publications.

Wang, Q. (2009) 'Gender, ethnicity, and self-employment: a multilevel analysis across US metropolitan areas', *Environment and Planning A* 41: 1979–1996.

Wang, Q. and Li, W. (2007) 'Entrepreneurship, ethnicity and local contexts: Hispanic entrepreneurs in three US southern metro areas', *GeoJournal* 68: 167–182.

Yeung, H. (2009) 'Transnationalizing entrepreneurship: a critical agenda for economic geography', *Progress in Human Geography* 33: 1–26.

Zhou, Y. (1998) 'Beyond ethnic enclaves: location strategies of Chinese producer service firms in Los Angeles', *Economic Geography* 74(3): 228–251.

Zook, M.A. (2004) 'The knowledge brokers: venture capitalists, tacit knowledge and regional development', *International Journal of Urban and Regional Research* 28: 621–641.

236

2.3 Accessibility

Aoyama, Y., Ratick, S.J., and Schwarz, G. (2006) 'Organizational dynamics of the U.S. logistics industry from an economic geography perspective', *Professional Geographer* 58(3): 327–340.

Aoyama, Y. and Ratick, S.J. (2007) 'Trust, transactions, and inter-firm relations in the U.S. logistics industry', *Economic Geography* 83(2): 159–180.

Bathelt, H. (2006) 'Geographies of production: growth regimes in spatial perspective 3 – toward a relational view of economic action and policy', *Progress in Human Geography* 30(2): 223–236.

Black, J. and Conroy, M. (1977) 'Accessibility measures and the social evaluation of urban structure', *Environment and Planning A* 9: 1013–1031.

Blumen, O. and Kellerman, A. (1990) 'Gender differences in commuting distance, residence, and employment location: metropolitan Haifa, 1972–1983', *The Professional Geographer* 42: 54–71.

Cairncross, F. (2001) *The Death of Distance: How the Communications Revolution is Changing Our Lives*. Cambridge, MA: Harvard Business School Press.

Cass, N., Shove, E. and Urry, J. (2005) 'Social exclusion, mobility and access', *The Sociological Review* 53(3): 539–555.

Christaller, W. (1966) *Central Places in Southern Germany*. An English translation of *Die zentralen Orte in Süddeutschland* by C.W. Baskin. Englewood Cliffs, NJ: Prentice Hall (originally published in 1933).

Cook, G., Pandit, N., Beaverstock, J., Taylor, P. and Pain, K. (2007) 'The role of location in knowledge creation and diffusion: evidence of centripetal and centrifugal forces in the City of London financial services agglomeration', *Environment and Planning A* 39: 1325–1345.

Crane, R. (2007) 'Is there a quiet revolution in women's travel? Revisiting the gender gap in commuting', *Journal of the American Planning Association* 73: 298–316.

Garrison, W.L. (ed.) (1959) *Studies in Highway Development and Geographic Change.* New York: Greenwood Press.

Goetz, A.R., Vowles, T.M. and Tierney, S. (2009) 'Bridging the qualitative-quantitative divide in transport geography', *The Professional Geographer* 61(3): 323–335.

Gilbert, M.R., Masucci, M., Homko, C. and Bove, A.A. (2008) 'Theorizing the digital divide: information and communication technology use frameworks among poor women using a telemedicine system', *Geoforum* 39(2): 912–925.

Hanson, S. and Johnston, I. (1985) 'Gender differences in worktrip length: explanations and implications,' *Urban Geography* 6: 193–219.

Hanson, S. and Pratt, G. (1991) 'Job search and the occupational segregation of women', *Annals of the Association of American Geographers* 81: 229–253.

Hiebert, D. (1999) 'Local geographies of labor market segmentation: Montreal, Toronto, and Vancouver, 1991', *Economic Geography* 75: 339–369.

Janelle, D. (2004) 'Impact of information technologies', in S. Hanson and G. Giuliano (eds), *The Geography of Urban Transportation*, 3rd edn. New York: Guilford Press. pp. 86–112.

Johnston-Anumonwo, I. (1997) 'Race, gender, and constrained work trips in Buffalo, NY, 1990', *The Professional Geographer* 49: 306–317.

Kain, J. (1968) 'Housing segregation, Negro employment, and metropolitan decentralization', *Quarterly Journal of Economics* 87: 175–197.

Kwan, M.-P. (1999) 'Gender and individual access to urban opportunities: a study using space-time measures', *The Professional Geographer* 51(2): 210–227.

Kwan, M.-P. and Weber, J. (2003) 'Individual accessibility revisited: implications for geographical analysis in the twenty-first century', *Geographical Analysis* 35(4): 1–13.

Leinbach, T.R., Bowen, J.T. (2004) 'Air cargo services and the electronics industry in Southeast Asia', *Journal of Economic Geography* 4: 299–321.

Madden, J. (1981) 'Why women work closer to home', *Urban Studies* 18: 181–194.

Malecki, E. and Moriset, B. (2008) *The Digital Economy. Business Organization, Production Processes and Regional Developments.* New York: Routledge.

Malecki, E. and Wei, H. (2009) 'A wired world: the evolving geography of submarine cables and the shift to Asia', *Annals of the Association of American Geographers*, 99: 360–382.

McLafferty, S. and Preston, V. (1991) 'Gender, race, and commuting among service sector workers', *The Professional Geographer* 43: 1–14.

Mokhtarian, P. and Meenakshisundaram, R. (1999) 'Beyond tele-substitution: disaggregate longitudinal structural equation modeling of communication impacts', *Transportation Research C* 7(1): 33–52.

Mokhtarian, P. (2003) 'Telecommunications and travel: the case for complementarity', *Journal of Industrial Ecology* 6: 43–57.

Murphy, J. T. (2006) 'Building trust in economic spaces', *Progress in Human Geography* 30(4): 427–450.

Parks, V. (2004a) 'Access to work: the effects of spatial and social accessibility on unemployment for native-born black and immigrant women in Los Angeles', *Economic Geography* 80(2): 141–172.

Parks, V. (2004b) 'The gendered connection between ethnic residential and labour-market segregation in Los Angeles', *Urban Geography* 25(7): 589–630.

Schafer, A. and Victor, D. (2000) 'The future of mobility of the world population', *Transportation Research A* 34(3): 171–205.

Schwanen, T. and Kwan, M.-P. (2008) 'The internet, mobile phone, and space-time constraints', *Geoforum* 39: 1362–1377.

Song Lee, B. and McDonald, J. (2003) 'Determinants of commuting time and distance for Seoul residents: the impact of family status on the commuting of women', *Urban Studies* 40: 1283–1302.

Transportation Research Board (2009) *Effects of Land Development Patterns on Motorized Travel, Energy, and CO_2 Emissions*. Washington, DC: Transportation Research Board.

Weber, A. (1929) *Theory of the Location of Industries*. An English translation of *Über den Standort der Industrien* by C.J. Friedrich. Chicago: University of Chicago Press (originally published in 1909).

Warf, B. (2001) 'Segueways into cyberspace: multiple geographies of the digital divide', *Environment and Planning B: Planning and Design* 28: 3–19.

Women and Geography Study Group (1984) *Geography and Gender*. London: Heinemann.

Wright, R. and Ellis, M. (2000) 'The ethnic and gender division of labour compared among immigrants to Los Angeles', *International Journal of Urban and Regional Research* 24: 583–600.

238

Industries and Regions in Economic Change

3.1 Industrial Location

Alonso, W. (1964) *Location and Land Use*. Cambridge, MA: Harvard University Press.

Barnes, T.J. (2004) 'The rise (and decline) of American regional science: lessons for the new economic geography?' *Journal of Economic Geography* 4(2): 107–129.

Barnes, T.J. (2000) 'The Space-Economy' Entry for Johnston et al. (eds), *The Dictionary of Human Geography*, 4th edn. pp. 773–774.

Brülhart, M. (1998) 'Economic geography, industry location and trade: The evidence', *The World Economy* 21(6): 775–801.

Dicken, P. (2007) *Global Shift: Mapping the Changing Contours of the World Economy*. 5th edn., London: Sage.

Frank, A.G. (1967) *Capitalism and Underdevelopment in Latin America*. New York: Monthly Review Press.

Holland, S. (1976). *Capital Versus the Regions*. New York: St. Martin's Press.

Hoover, E.M. (1967) 'Some programmed models of Industry Location', *Land Economics* 43: 303–311.

Isard, W. (1956) *Location and Space-Economy*. New York: Wiley.

Krugman, P.R. (1991) *Geography and Trade*. Cambridge, MA: MIT Press.

Kuhn, H.W. and Kuenne, R.E. (1962) 'An efficient algorithm for the numerical solution of the generalized Weber problem in space economics', *Journal of Regional Science* 4: 21–33.

Lösch, A. (1954 [1940]) *The Economics of Location*. An English translation of *Die räumliche Ordnung der Wirtschaft* by W.H. Woglom. New Haven, CT: Yale University Press (originally published in 1940).

Massey, D. (1979) 'In what sense a regional problem?', *Regional Studies* 13(2): 233–243.

Pred, A. (1967) 'Behavior and location: foundations for a geographic and dynamic location theory, Part 1', *Lund Studies in Geography*, Series B, 27.

Smith, D.M. (1981) *Industrial Location: An Economic Geographical Analysis*, 2nd edn. New York: John Wiley & Sons.

Storper, M. and Scott, A.J. (2009) 'Rethinking human capital, creativity and urban growth', *Journal of Economic Geography* 9: 147–167.

Storper, M. and Walker, R.A. (1989) *The Capitalist Imperative: Territory, Technology, and Industrial Growth*. Oxford: Basil Blackwell.

Taylor, M.J. and Thrift, N.J. (1982) The *Geography of Multinationals*. London: Croom Helm.

Wallerstein, I. (1979) *The Capitalist World-Economy*. Cambridge: Cambridge University Press.

Weber, A. (1929 [1909]) *Theory of the Location of Industries*. An English translation of *Über den Standort der Industrien* by C.J. Friedrich. Chicago: University of Chicago Press (originally published in 1909).

3.2 Industrial Clusters

Amin, A. and Cohendet, P. (2004) *Architecture of Knowledge: Firms, Capabilities and Communities*. Oxford: Oxford University Press.

Amin, A. and Thrift, N. (1992) 'Neo-Marshallian nodes in global networks', *International Journal of Urban and Regional Research* 16: 571–587.

Asheim, B.T. and Coenen, L. (2005) 'Knowledge bases and regional innovation systems: comparing Nordic clusters', *Research Policy* 34(8): 1173–1190.

Audretsch, D. and Feldman, M.P. (1996) 'R&D spillovers and the geography of innovation and production', *American Economic Review* 86(3): 630–640.

Aydalot, P. (ed.) (1986) *Milieux Innovateurs en Europe*. Paris: GREMI.

Bagnasco, A. (1977) *Tre Italie: La Problematica Territoriale Dello Sviluppo Italiano*. Bologna: Il Mulino.

Bathelt, H. (2005) 'Geographies of production: growth regimes in spatial perspective (II) – knowledge creation and growth in clusters', *Progress in Human Geography* 29(2): 204–216.

Bathelt, H., Malmberg, A. and Maskell, P. (2004) 'Clusters and knowledge: local buzz, global pipelines and the process of knowledge creation', *Progress in Human Geography* 28(1): 31–56.

Brown, J. and Duguid, P. (1991) 'Organizational learning and communities of practice: Toward a unified view of working, learning, and innovation', *Organization Science* 2: 40–57.

Brusco, S. (1982) 'The Emilian model: productive decentralisation and local integration', *Cambridge Journal of Economics* 6(2): 167–184.

Castells, M. and Hall, P. (1994) *Technopoles of the World: The Making of Twenty-First-Century Industrial Complexes*. London: Routledge.

References

Clark, G.L. (2002) 'London in the European financial services industry: locational advantage and product complementarities', *Journal of Economic Geography* 2(4): 433–453.

Cooke, P. and Morgan, K. (1994) 'The regional innovation system in Baden-Württemberg', *International Journal of Technology Management* 9: 394–429.

Feldman, M. P. (1994) 'Knowledge complementarity and innovation', *Small Business Economics,* 6(5): 363–372.

Feldman, M.P. and Audretsch, D.B. (1999) 'Innovation in cities: science-based diversity, specialization and localized competition', *European Economic Review* 43: 409–429.

Florida, R. (2002) *The Rise of the Creative Class: And How It's Transforming Work, Leisure, Community and Everyday Life.* New York: Basic Books.

Friedman, T.L. (2005) *The World is Flat: A Brief History of the Twenty-first Century.* New York: Farrar, Straus and Giroux.

Glaeser, E.L., H.D. Kallal, J.A. Scheinkman and Shleifer, A. (1992) 'Growth of cities', *Journal of Political Economy* 100: 1126–1152.

Granovetter, M. (1985) 'Economic action and social structure: the problem of embeddedness', *American Journal of Sociology* 91(3): 480–510.

Harrison, B. (1992) 'Industrial districts: old wine in new bottles', *Regional Studies* 26(5): 469–483.

Harrison, B., Kelley, M.R. and Jon, G. (1996) 'Innovative firm behavior and local milieu: exploring the intersection of agglomeration, firm effects, and technological change', *Economic Geography* 72(3): 233–258.

Hoover, E.M. (1948) *The Location of Economic Activity.* New York: McGraw Hill.

Jacobs, J. (1969) *The Economy of Cities.* New York: Vintage.

Leamer, E. and Storper, M. (2001) 'The economic geography of the internet age', *Journal of International Business Studies* 32(4): 641–666.

Markusen, A. (1996) 'Sticky places in slippery space: a typology of industrial districts', *Economic Geography* 72(3): 293–313.

Marshall, A. (1920) *Principles of Economics.* First published in 1890. London: Macmillan.

Martin, R. and Sunley, P. (2003) 'Deconstructing clusters: chaotic concept or policy panacea?', *Journal of Economic Geography* 3(1): 5–35.

Piore, M.J. and Sabel, C.F. (1984) *The Second Industrial Divide: Possibilities for Prosperity.* New York: Basic Books.

Porter, M.E. (2000) 'Location, competition, and economic development: local clusters in a global economy', *Economic Development Quarterly* 14(1): 15–34.

Ratti, R. (1992) 'Eléments de théorie économique des effets frontiers et de politique de développement regional Exemplification d'après le cas des agglomérations de frontière Suisses', *Revue Suisse d'Economie Politique et de Statistique* 128(3): 325–338.

Richardson, Harry W. and Richardson, M. (1975) 'The relevance of growth center strategies to Latin America', *Economic Geography* 51(2): 163–178.

Russo, M. (1985) 'Technical change and the industrial district: the role of interfirm relations in the growth and transformation of ceramic tile production in Italy', *Research Policy* 14(6): 329–343.

Saxenian, A. (1994) *Regional Advantage: Culture and Competition in Silicon Valley and Route 128.* Cambridge, MA: Harvard University Press.

Scott, A. (1988) *New Industrial Spaces: Flexible Production Organization and Regional Development in North America and Western Europe.* London: Pion.

Scott, A. (2005) *On Hollywood: The Place, the Industry.* Princeton, NJ: Princeton University Press.

Simmel, G. (1950) *The Sociology of Georg Simmel.* Compiled and translated by Kurt Wolff. Glencoe, IL: The Free Press.

Storper, M. (1995) 'The resurgence of regional economies, ten years later: The region as a nexus of untraded interdependencies', *European Urban and Regional Studies* 2: 191–221.

Storper, M. (1997) *The Regional World: Territorial Development in a Global Economy.* New York: Guilford.

Storper, M. (2009) 'Roepke lecture in economic geography – regional context and global trade', *Economic Geography* 85(1) (January): 1–22.

Storper, M. and Venables, A.J. (2004) 'Buzz: face-to-face contact and the urban economy', *Journal of Economic Geography* 4(4): 351–370.

Van Oort, F. (2002) 'Innovation and agglomeration economies in the Netherlands', *Tijdschrift voor economische en sociale geografie* 93(3): 344–360.

Wenger, E. (1999) *Communities of Practice: Learning, Meaning and Identity.* Cambridge: Cambridge University Press.

Young, A. (1928) 'Increasing returns and economic progress', *Economic Journal* 38: 527–542.

3.3 Regional Disparity

Alonso, W. (1968) 'Urban and regional imbalances in economic development', *Economic Development and Cultural Change* 17(1): 1–14.

Arthur, W.B. (1989) 'Competing technologies, increasing returns, and lock-in by historical events', *Economic Journal* 99: 116–131.

Castells, M. and Hall, P.G. (1994) *Technopoles of the World: The Making of Twenty-First-Century Industrial Complexes.* London: Routledge.

Clark, C. (1940) *The Conditions of Economic Progress.* London: Macmillan.

Combes, P.-P., Mayer, T. and Thisse, J.-F. (2008) *Economic Geography: The Integration of Regions and Nations.* Princeton: Princeton University Press.

Harrison, B. and Bluestone, B. (1988) *The Great U-Turn: Corporate Restructuring and the Polarizing of America.* New York: Basic Books.

Hirschman, A.O. (1958) *The Strategy of Economic Development.* New Haven, CT: Yale University Press.

Holland, S. (1976) *Capital Versus the Regions.* New York: St. Martin's Press.

Hoover, E.M. and Fisher, J. (1949) 'Research in regional economic growth', Chapter V in Universities-National Bureau Committee for Economic Research (ed.) *Problems in the Study of Economic Growth.* New York: National Bureau of Economic Research.

Hudson, R. (2007) 'Regions and regional uneven development forever? Some reflective comments upon theory and practice', *Regional Studies* 41(9): 1149–1160.

Jacobs, J. (1969) *The Economy of Cities.* New York: Vintage.

Krugman, P. (1991) 'Increasing returns and economic geography', *The Journal of Political Economy* 99(3): 483–499.

Kuklinski, A.R. (ed.) (1972) *Growth Poles and Growth Centres in Regional Planning.* The Hague: Mouton.

Kuznets, S. (1955) 'Economic growth and income inequality', *American Economic Review* 65(March): 1–28.

Lanaspa, L.F. and Fernando, S. (2001) 'Multiple equilibria, stability, and asymmetries in Krugman's core–periphery model', *Papers in Regional Science* 80: 425–438.

Lanaspa, L.F. and Sanz, F. (2001) 'Multiple equilibria, stability, and asymmetries in Krugman's core-periphery model', *Papers in Regional Science* 80, 425–438.

Markusen, A.R., Hall, P., Campbell, S. and Deitrick, S. (1991) *The Rise of the Gunbelt: The Military Remapping of Industrial America*. New York: Oxford University Press.

Myrdal, G. (1957) *Economic Theory and Under-developed Regions*. London: Gerald Duckworth.

North, D.C. (1955) 'Location theory and regional economic growth', *The Journal of Political Economy* 63(3): 243–258.

North, D.C. (1956) 'A Reply', *The Journal of Political Economy* 64(2): 165–168.

Ohlin, B. (1933) *Interregional and International Trade*. Cambridge, MA: Harvard University Press.

Perroux, F. (1950) 'Economic space, theory and applications', *Quarterly Journal of Economics* 64(1): 89–104.

Pike, A., Rodríguez-Pose, A. and Tomaney, J. (2006) *Local and Regional Development*. London: Routledge.

Richardson, H.W. and M. Richardson (1975) 'The relevance of growth center strategies to Latin America', *Economic Geography* 51(2): 163–178.

Tiebout, C.M. (1956a) 'Exports and regional economic growth', *The Journal of Political Economy* 64(2): 160–164.

Tiebout, C.M. (1956b) 'Exports and regional economic growth: rejoinder', *The Journal of Political Economy* 64(2): 169.

World Bank (2009) *World Development Report: Reshaping Economic Geography*. Washington, DC: The World Bank.

3.4 Post-Fordism

Aglietta, M. (1979 [1976]) *A Theory of Capitalist Regulation: The US Experience*. An English translation of *Régulation et Crises du Capitalisme*. London: New Left Books (originally published in 1976).

Amin, A. and K. Robins (1990) 'The re-emergence of regional economies? The mythical geography of flexible accumulation', *Environment and Planning D: Society and Space* 8(1): 7–34.

Bagnasco, A. (1977) *Tre Italie: La Problematica Territoriale Dello Sviluppo Italiano*. Bologna: Il Mulino.

Best, M. (1990) *The New Competition*. Cambridge, MA: Harvard University Press.

Bluestone, B. and Harrison, B. (1982) *Deindustrialization of America: Plant Closings, Community Abandonment, and the Dismantling of Basic Industries*. New York: Basic Books.

Boyer, R. (1979) 'Wage formation in historical perspective: The French experience', *Cambridge Journal of Economics* 3(2): 99–118.

Boyer, R. (1990) *The Regulation School Approach: A Critical Introduction*. New York: Columbia University Press.

Brusco, S. (1982) 'The Emilian model: productive decentralisation and local integration', *Cambridge Journal of Economics* 6(2): 167–184.

Chandler, A.D. (1977) *The Visible Hand: The Managerial Revolution in American Business*. Cambridge, MA: Harvard University Press.

References

Christopherson, S. and Storper, M. (1986) 'The city as studio; the world as back lot: the impact of vertical disintegration on the motion picture industry', *Environment and Planning D* 4(3): 305–320.

Cooke, P. and Morgan, K. (1994) 'The regional innovation system in Baden-Württemberg', *International Journal of Technology Management* 9: 394–429.

Coriat, B. (1979) *L'atelier et le chronometer: Essai sur le taylorisme, le fordisme et la production de masse*. Paris: Christian Bourgois Editeur.

Dunford, M. (1990) 'Theories of regulation', *Environment and Planning D* 8(3): 297–321.

Fujita, K. and Hill, R.C. (1993) 'Toyota city: industrial organization and the local state in Japan', in K. Fujita and R.C. Hill (eds) *Japanese Cities in the World Economy*. Philadelphia, NJ: Temple University Press. pp. 175–199.

Gertler, M. (1988) 'The limits to flexibility: comments on the Post-Fordist vision of production and its geography', *Transactions of the Institute of British Geographers* 13(4): 419–432.

Gertler, M. (1989) 'Resurrecting flexibility? A reply to Schoenberger', *Transactions of the Institute of British Geographers* 14(1): 109–112.

Harrison, B. (1994) *Lean and Mean: The Changing Landscape of Corporate Power in the Age of Flexibility*. New York: Basic Books.

Herrigel, G. (1996) 'Crisis in German decentralized production: unexpected rigidity and the challenge of an alternative form of flexible organization in Baden Wurttemberg', *European Urban and Regional Studies* 3(1): 33–52.

Jessop, B. (1990) 'Regulation theories in retrospect and prospect', *Economy and Society* 19(2): 153–216.

Kenney, M. (ed.) (2000) *Understanding Silicon Valley: The Anatomy of an Entrepreneurial Region*. Stanford, CA: Stanford Business Books.

Lipietz, A. (1986) 'New tendencies in the international division of labour: regimes of accumulation and modes of regulation', in A.J. Scott and Storper, M. (eds) *Production, Work, Territory: The Geographical Anatomy of Industrial Capitalism*. Boston: Allen & Unwin. pp. 16–40.

Peck, J. and Theodore, N. (1998) 'The business of contingent work: growth and restructuring in Chicago's temporary employment industry', *Work, Employment and Society* 12(4): 655–674.

Piore, M.J. and Sabel, C.F. (1984) *The Second Industrial Divide: Possibilities for Prosperity*. New York: Basic Books.

Russo, M. (1985) 'Technical change and the industrial district: the role of interfirm relations in the growth and transformation of ceramic tile production in Italy', *Research Policy* 14(6): 329–343.

Rutherford, T. and Gertler, M. (2002) 'Labour in lean times: geography, scale and the national trajectories of workplace change', *Transactions of the Institute of British Geographers* 27: 195–212.

Sabel, C., Herrigel, G., Deeg, R. and Kazis, R. (1989) 'Regional prosperities compared: Baden-Württemberg and Massachusetts in the 1980s', *Economy and Society* 18 (4): 374–404.

Saxenian, A. (1985) 'Silicon Valley and Route 128: regional prototypes or historic exceptions?', in M. Castells (ed.) *High Technology, Space and Society*. Beverley Hills, CA: Sage Publications.

Saxenian, A. (1994) *Regional Advantage: Culture and Competition in Silicon Valley and Route 128*. Cambridge, MA: Harvard University Press.

Schoenberger, E. (1988) 'From Fordism to flexible accumulation: technology, competitive strategies, and international location', *Environment and Planning D* 6(3): 245–262.

Schoenberger, E. (1989) 'Thinking about flexibility: a response to Gertler', *Transactions of the Institute of British Geographers* 14(1): 98–108.

Storper, M. and Christopherson, S. (1987) 'Flexible specialization and regional industrial agglomerations – the case of the United-States motion-picture industry', *Annals of the Association of American Geographers* 77(1): 104–117.

Taylor, F.W. (1911) *Principles of Scientific Management.* New York and London: Harper & Brothers.

Womack, J.P., Jones, D.T. and Roos, D. (1990) *The Machine that Changed the World: The Story of Lean Production.* New York: HarperPerennial.

4 Global Economic Geographies

4.1 Core–Periphery

Arrighi, G. (2002) 'The African crisis: world systemic and regional aspects', *New Left Review* 15: 5–36.

Arrighi, G. (2007) *Adam Smith in Beijing: Lineages of the Twenty-first Century.* London: Verso.

Auty, R.M. (1993) *Sustaining Development in Mineral Economies: The Resource Curse Thesis.* London: Routledge.

Baran, P. (1957) *The Political Economy of Growth.* New York: Monthly Review Press.

Barton, J.R., Gwynne, R.N. and Murray, W.E. (2007) 'Competition and co-operation in the semi-periphery: closer economic partnership and sectoral transformations in Chile and New Zealand', *Geographical Journal* 173(3): 224–241.

Castells, M. (1996) *The Rise of the Network Society: The Information Age: Economy, Society and Culture Vol. I.* Cambridge, MA: Blackwell.

Castells, M. (1998) *End of the Millennium: The Information Age: Economy, Society and Culture Vol. III.* Cambridge, MA: Blackwell.

Emmanuel, A. (1972) *Unequal Exchange: A Study of the Imperialism of Trade.* London: New Left Books.

Fage, J.D., Roberts, A.D. and Oliver, R.A. (1986) *The Cambridge History of Africa.* Cambridge: Cambridge University Press.

Frank, A.G. (1966) 'The development of underdevelopment', *Monthly Review* 18: 17–31.

Frank, A.G. (1998) *Reorient: Global Economy in the Asian Age.* Berkeley, CA: University of California Press.

Freudenburg, W. (1992) 'Addictive economies: extractive industries and vulnerable localities in a changing world economy', *Rural Sociology* 57: 305–332.

Gilbert, M.R., Masucci, M., Homko, C. and Bove, A.A. (2008) 'Theorizing the digital divide: information and communication technology use frameworks among poor women using a telemedicine system', *Geoforum* 39(2): 912–925.

Gwynne, R.N., Klak, T. and Shaw, D.J.B. (2003) *Alternative Capitalisms: Geographies of Emerging Regions.* London: Arnold.

References

Hall, T.D. (2000) 'World-systems analysis: A small sample from a large universe', in T.D. Hall (ed.) *A World-Systems Reader: New Perspectives on Gender, Urbanism, Cultures, Indigenous Peoples, and Ecology*. Lanham, MD: Rowman & Littlefield. pp. 3–27.

Harvey, D. (2006) *Spaces of Global Capitalism: Towards a Theory of Uneven Geographical Development*. London: Verso.

Hayter, R., Barnes, T.J. and Bradshaw, M.J. (2003) 'Relocating resource peripheries to the core of economic geography's theorizing: Rationale and agenda', *Area* 35(1): 15–23.

Kellerman, A. (2002) *The Internet on Earth: A Geography of Information*. New York: Wiley.

Klak, T. (ed.) (1998) *Globalization and Neoliberalism: The Caribbean Context*. Lanham, MD: Rowman and Littlefield.

Makki, F. (2004) 'The empire of capital and the remaking of centre–periphery relations', *Third World Quarterly* 25(1): 149–168.

Pain, K. (2008) 'Examining 'core–periphery' relationships in a global city-region: the case of London and South East England', *Regional Studies* 42(8): 1161–1172.

Rosser, A. (2007) 'Escaping the resource curse: The case of Indonesia', *Journal of Contemporary Asia* 37(1): 38–58.

Rostow, W.W. (1960) *The Stages of Economic Growth: A Non-Communist Manifesto*. Cambridge: Cambridge University Press.

Straussfogel, D. (1997) 'World-systems theory: toward a heuristic and pedagogic conceptual tool', *Economic Geography* 73(1): 118–130.

Taylor, P.J. and Flint, C. (2000) *Political Geography: World-system, Nation-State and Locality*. London: Longman.

Terlouw, K. (2009) 'Transnational regional development in the Netherlands and Northwest Germany, 1500–2000', *Journal of Historical Geography* 35(1): 26–43.

Wallerstein, I. (1974) *The Modern World-System: Capitalist Agriculture and the Origins of the European World-Economy in the Sixteenth Century*. New York: Academic Press.

Williamson, J. (2004) 'The Washington Consensus as policy prescription for development', World Bank Practitioners of Development lecture given on 13 January 2004. Transcript accessed from the Institute for International Economics webpage on 8 June 2009: http://www.iie.com/publications/papers/williamson0204.pdf.

World Bank (1993) *The East Asian Miracle: Economic Growth and Public Policy*. New York: Oxford University Press.

245

4.2 Globalization

Amin, A. (2002) 'Spatialities of globalisation', *Environment and Planning A* 34(3): 385–399.

Amin, A. and Graham, S. (1997) 'The ordinary city', *Transactions of the Institute of British Geographers* 22(4): 411–429.

Amin, A. and Thrift, N. (1993) 'Globalization, institutional thickness, and local prospects', *Revue d'Economie Regionale et Urbaine* 3: 405–427.

Appadurai, A. (1990) 'Disjuncture and difference in the global cultural economy', *Theory, Culture & Society* 7(2): 295–310.

References

Appadurai, A. (1996) *Modernity at Large: Cultural Dimensions of Globalization.* Minneapolis, MN: University of Minnesota Press.

Bardhan, A.D. and Howe, D.K. (2001) 'Globalization and restructuring during downturns: a case study of California', *Growth and Change* 32(2): 217–235.

Bhagwati, J. (2004) *In Defense of Globalization.* New York: Oxford University Press.

Brenner, N. (2000) 'The urban question as a scale question: reflections on Henri Lefebvre, urban theory and the politics of scale', *International Journal of Urban and Regional Research* 24(2): 361–378.

Bridge, G. (2002) 'Grounding globalization: the prospects and perils of linking economic processes of globalization to environmental outcomes', *Economic Geography* 78(3): 361–386.

Castells, M. (1996) *The Rise of the Network Society: The Information Age: Economy, Society and Culture Vol. I.* Cambridge, MA: Blackwell.

Cox, K.R. (2008) 'Globalization, uneven development and capital: reflections on reading Thomas Friedman's *The World Is Flat*', *Cambridge Journal of Regions, Economy and Society* 1(3): pp. 389–410.

Dicken, P. (1994) 'Global–local tensions: firms and states in the global space-economy', *Economic Geography* 70(2): 101–128.

Dicken, P. (2004) 'Geographers and 'globalization': (yet) another missed boat?', *Transactions of the Institute of British Geographers* 29(1): 5–26.

Dicken, P. (2007) *Global Shift. Mapping the Changing Contours of the World Economy,* 5th edn. London: Sage.

Friedman, T.L. (2005) *The World Is Flat: A Brief History of the Twenty-first Century.* New York: Farrar, Straus and Giroux.

Fröbel, F., Heinrichs, J. and Kreye, O. (1978) 'The world market for labour and the world market for industrial sites', *Journal of Economic Issues* 12(4): 843–858.

Gereffi, G. (1999) 'International trade and industrial upgrading in the apparel commodity chain,' *Journal of International Economics* 48: 37–70.

Grant, R. and Nijman, J. (2004) 'The rescaling of uneven development in Ghana and India', *Tijdschrift voor Economische en Sociale Geografie* 95(5): 467–481.

Held, D. and McGrew, A. (2007) *Globalization/Anti-Globalization: Beyond the Great Divide,* 2nd edn. Cambridge: Polity Press.

Hess, M. (2004) '"Spatial" relationships? Towards a reconceptualization of embeddedness', *Progress in Human Geography* 28(2): 165–186.

Hymer, S. (1976) *The International Operations of National Firms: A Study of Foreign Direct Investment.* Cambridge, MA: MIT Press.

Jessop, B. (1999) 'Reflections on globalisation and its (il)logic(s)', in K. Olds, P. Dicken, P.F. Kelly, L. Kong and H.W.-C. Yeung (eds) *Globalisation in the Asia-Pacific: Contested Territories.* London: Routledge. pp. 19–38.

Jones, R.C. (1998) 'Remittances and inequality: a question of migration stage and geographic scale', *Economic Geography* 74(1): 8–25.

Kelly, P.F. (1999) 'The geographies and politics of globalization', *Progress in Human Geography* 23(3): 379–400.

Marston, S.A., Jones III, J.P. and Woodward, K. (2005) 'Human geography without scale', *Transactions of the Institute of British Geographers* 30(4): 416–432.

Maskell, P. and Malmberg, A. (1999) 'Localized learning and industrial competitiveness', *Cambridge Journal of Economics* 23: 167–185.

References

Mitchell, K. (1995) 'Flexible circulation in the Pacific Rim: capitalisms in cultural context', *Economic Geography* 71(4): 364–382.

Mohan, G. and Zack-Williams, A.B. (2002) 'Globalisation from below: conceptualising the role of the African Diasporas in Africa's development', *Review of African Political Economy* 29(92): 211–236.

Nagar, R., Lawson, V., McDowell, L. and Hanson, S. (2002) 'Locating globalization: feminist (re)readings of the subjects and spaces of globalization', *Economic Geography* 78(3): 257–284.

Reinert, K.A. (2007) 'Ethiopia in the world economy: trade, private capital flows, and migration', *Africa Today* 53(3): 65–89.

Rigg, J., Bebbington, A., Gough, K.V., Bryceson, D.F., Agergaard, J., Fold, N. and Tacoli, C. (2009) '*The World Development Report 2009* "reshapes economic geography": geographical reflections', *Transactions of the Institute of British Geographers* 34(2): 128–136.

Rosen, E.I. (2002) *Making Sweatshops: The Globalization of the U.S. Apparel Industry*. Berkeley, CA: University of California Press.

Saxenian, A.L. and Hsu, J.-Y. (2001) 'The Silicon Valley–Hsinchu connection: technical communities and industrial upgrading', *Industrial and Corporate Change* 10(4): 893–920.

Scott, A.J. (2006) 'The changing global geography of low-technology, labour-intensive industry: clothing, footwear, and furniture', *World Development* 34(9): 1517–1536.

Sheppard, E. (2002) 'The spaces and times of globalization: place, scale, networks, and positionality', *Economic Geography* 78(3): 307–330.

Sklar, L. (2002) *Globalization Capitalism and its Alternatives*. Oxford: Oxford University Press.

Stiglitz, J. (2002) *Globalization and Its Discontents*. New York: W.W. Norton.

Storper, M. (1992) 'The limits to globalization: technology districts and international trade', *Economic Geography* 68(1): 60–93.

Swyngedouw, E.A. (1992) 'Territorial organization and the space/technology nexus', *Transactions of the Institute of British Geographers* 17(4): 417–433.

Swyngedouw, E.A. (1997) 'Excluding the other: the production of scale and scaled politics', in R. Lee and J. Wills (eds) *Geographies of Economies*. London: Arnold. pp. 167–176.

Taylor, M.J. and Thrift, N.J. (1982) 'Models of corporate development and the multinational corporation', in M.J. Taylor and N.J. Thrift (eds) *The Geography of Multinationals: Studies in the Spatial Development and Economic Consequences of Multinational Corporations*. New York: St. Martin's Press. pp. 14–32.

Tokatli, N. (2008) 'Global sourcing: Insights from the global clothing industry – the case of Zara, a fast fashion retailer', *Journal of Economic Geography* 8(1): 21–38.

Vernon, R. (1966) 'International investment and international trade in the product cycle', *The Quarterly Journal of Economics* 80(2): 190–207.

Wade, R.H. (2004) 'Is globalization reducing poverty and inequality?', *World Development*, 32(4): 567–589.

Wong, M. (2006) 'The gendered politics of remittances in Ghanaian transnational families', *Economic Geography* 82(4): 355–382.

World Bank (2009) *2009 World Development Report: Reshaping Economic Geography*. Washington: World Bank.

Yeung, H.W.C. (1998) 'Capital, state and space: contesting the borderless world', *Transactions of the Institute of British Geographers* 23: 291–309.

4.3 Circuits of Capital

Arrighi, G. (2002) 'The African crisis: world systemic and regional aspects', *New Left Review* 15: 5–36.

Bartelt, D.W. (1997) 'Urban housing in an era of global capital', *Annals of the American Academy of Political and Social Science* 551: 121–136.

Bello, W. (2006) 'The capitalist conjuncture: over-accumulation, financial crises, and the retreat from globalisation', *Third World Quarterly* 27(8): 1345–1367.

Brenner, N. (2001) 'The limits to scale? Methodological reflections on scalar structuration', *Progress in Human Geography* 25(4): 591–614.

Christophers, B. (2006) 'Circuits of capital, genealogy, and television geographies', *Antipode* 38(5): 930–952.

Desai, M. (1979) *Marxian Economics*. Oxford: Blackwell.

Foot, S.P.H. and Webber, M. (1990) 'State, class and international capital 2: the development of the Brasilian steel industry', *Antipode* 22(3): 233–251.

Gibson-Graham, J.K. (2008) 'Diverse economies: Performative practices for "other worlds"', *Progress in Human Geography* 32(5): 613–632.

Glassman, J. (2001) 'Economic crisis in Asia: the case of Thailand', *Economic Geography* 77(2): 122–147.

Glassman, J. (2007) 'Recovering from crisis: the case of Thailand's spatial fix', *Economic Geography* 83(4): 349–370.

Harvey, D. (1982) *The Limits to Capital*. London: Blackwell.

Harvey, D. (1989) *The Urban Experience*. Baltimore, MD: Johns Hopkins University Press.

Harvey, D. (2001) *Spaces of Capital: Towards a Critical Geography*. New York: Routledge.

Hudson, R. (2004) 'Conceptualizing economies and their geographies: spaces, flows and circuits', *Progress in Human Geography* 28(4): 447–471.

Hung, H.F. (2008) 'Rise of China and the global overaccumulation crisis', *Review of the International Political Economy* 15(2): 149–179.

Jessop, B. (2000) 'The crisis of the national spatio-temporal fix and the tendential ecological dominance of globalizing capitalism', *International Journal of Urban and Regional Research* 24(2): 323–360.

Jones, M. and Ward, K. (2004) 'Capitalist development and crisis theory: towards a "fourth cut"', *Antipode* 36(3): 497–511.

King, R., Dalipaj, M. and Mai, N. (2006) 'Gendering migration and remittances: evidence from London and northern Albania', *Population, Space and Place* 12(6): 409–434.

Lee, R. (2002) '"Nice maps, shame about the theory"? Thinking geographically about the economic', *Progress in Human Geography* 26(3): 333–355.

Lee, R. (2006) 'The ordinary economy: tangled up in values and geography', *Transactions of the Institute of British Geographers* 31(4): 413–432.

Leyshon, A. and Thrift, N. (1996) *Money/Space: Geographies of Monetary Transformation*. London: Routledge.

Marx, K. (1967) *Capital*. Volumes I–III, New York: International Publishers.

McMichael, P. (1991) 'Slavery in capitalism: the rise and demise of the U.S. antebellum cotton culture', *Theory and Society* 20(3): 321–349.

Peterson, V.S. (2003) *A Critical Rewriting of Global Political Economy: Integrating Reproductive, Productive, and Virtual Economies.* London: Routledge.

Roberts, S.M. (1995) 'Small place, big money: the Cayman Islands and the international financial system', *Economic Geography* 71(3): 237–256.

Schoenberger, E. (2004) 'The spatial fix revisited', *Antipode* 36(3): 427–433.

Smith, N. (1990) *Uneven Development: Nature, Capital and the Production of Space.* Oxford: Basil Blackwell.

Smith, N. (2002) 'New globalism, new urbanism: gentrification as global urban strategy', *Antipode* 34(3): 427–450.

Swyngedouw, E. (1997) 'Excluding the other: the production of scale and scaled politics', in R. Lee and J. Wills (eds) *Geographies of Economies.* London: Arnold. pp. 167–176.

Warf, B. (2002) 'Tailored for Panama: offshore banking at the crossroads of the Americas', *Geografiska Annaler Series B: Human Geography* 84(1): 33–47.

Wilson, B.M. (2005) 'Race in commodity exchange and consumption: separate but equal', *Annals of the Association of American Geographers* 95(3): 587–606.

Wyly, E.K., Atia, M. and Hammel, D.J. (2004) 'Has mortgage capital found an inner-city spatial fix?', *Housing Policy Debate* 15(3): 623–685.

4.4 Global Value Chains

Barrientos, S., Dolan, C. and Tallontire, A. (2003) 'A gendered value chain approach to codes of conduct in African horticulture', *World Development* 31(9): 1511–1526.

Coe, N.M., Dicken, P. and Hess, M. (2008) 'Global production networks: realizing the potential', *Journal of Economic Geography* 8(3): 271–295.

Coe, N.M., Hess, M., Yeung, H.W., Dicken, P. and Henderson, J. (2004) '"Globalizing" regional development: a global production networks perspective', *Transactions of the Institute of British Geographers* 29(4): 468–484.

Dicken, P., Kelly, P.F., Olds, K. and Yeung, H.W. (2001) 'Chains and networks, territories and scales: towards a relational framework for analyzing the global economy', *Global Networks* 1(2): 89–112.

Fold, N. (2002) 'Lead firms and competition in "Bi-polar" commodity chains: grinders and branders in the global cocoa-chocolate industry', *Journal of Agrarian Change* 2(2): 228–247.

Gereffi, G. (1994) 'The organization of buyer-driven global commodity chains: how U.S. retailers shape overseas production networks', in G. Gereffi and M. Koreniewicz (eds) *Commodity Chains and Global Capitalism.* Westport, CT: Praeger. pp. 95–122.

Gereffi, G. (1995) 'Global production systems and Third World development', in B. Stallings (ed.) *Global Change, Regional Response: The New International Context of Development.* Cambridge: Cambridge University Press. pp. 100–142.

Gereffi, G. (1999) 'International trade and industrial upgrading in the apparel commodity chain', *Journal of International Economics* 48(1): 37–70.

Gereffi, G. and Korzeniewicz, M. (eds) (1994) *Commodity Chains and Global Capitalism.* Westport, CT: Praeger.

References

Gereffi, G., Humphrey, J. and Sturgeon, T. (2005) 'The governance of global value chains', *Review of International Political Economy* 12(1): 78–104.

Gereffi, G., Koreniewicz, M. and Koreniewicz, R.P. (1994) 'Introduction: global commodity chains', in G. Gereffi and M. Koreniewicz (eds) *Commodity Chains and Global Capitalism*. Westport, CT: Praeger. pp. 1–14.

Gibbon, P. (2003) 'The African growth and opportunity act and the global commodity chain for clothing', *World Development* 31(11): 1809–1827.

Gibbon, P. and Ponte, S. (2005) *Trading Down: Africa, Value Chains, and the Global Economy*. Philadelphia, NJ: Temple University Press.

Giuliani, E., Pietrobelli, C. and Rabellotti, R. (2005) 'Upgrading in global value chains: lessons from Latin American clusters', *World Development* 33(4): 549–573.

Henderson, J., Dicken, P., Hess, M., Coe, N.M. and Yeung, H.W. (2002) 'Global production networks and the analysis of economic development', *Review of International Political Economy* 9(3): 436–464.

Hilson, G. (2008) '"Fair trade gold": antecedents, prospects and challenges', *Geoforum* 39(1): 386–400.

Hopkins, T.K. and Wallerstein, I. (1986) 'Commodity chains in the world-economy prior to 1800', *Review* 10(1): 157–170.

Hudson, R. (2008) 'Cultural political economy meets global production networks: a productive meeting?', *Journal of Economic Geography* 8(3): 421–440.

Humphrey, J. (2003) 'Globalization and supply chain networks: the auto industry in Brazil and India', *Global Networks* 3(2): 121–141.

Humphrey, J. and Schmitz, H. (2000) 'Governance and upgrading: linking industrial cluster and global value chain research', *IDS Working Paper 120*, Brighton: Institute of Development Studies at the University of Sussex.

Humphrey, J. and Schmitz, H. (2002) 'Developing country firms in the world economy: governance and upgrading in global value chains', INEF Report Heft 61/2002, Institut für Entwicklung und Frieden der Gerhard-Mercator-Universität Duisburg.

Izushi, H. (1997) 'Conflict between two industrial networks: technological adaptation and inter-firm relationships in the ceramics industry in Seto, Japan', *Regional Studies* 31(2): 117–129.

Morris, M. and Dunne, N. (2004) 'Driving environmental certification: its impact on the furniture and timber products value chain in South Africa', *Geoforum* 35(2): 251–266.

Neilson, J. (2008) 'Global private regulation and value-chain restructuring in Indonesian smallholder coffee systems', *World Development* 36(9): 1607–1622.

Neumayer, E. and Perkins, R. (2005) 'Uneven geographies of organizational practice: explaining the cross-national transfer and diffusion of ISO 9000', *Economic Geography* 81(3): 237–259.

Okada, A. (2004) 'Skills development and inter-firm learning linkages under globalization: lessons from the Indian automobile industry', *World Development* 32(7): 1265–1288.

Ouma, S. (2010) 'Global standards, local realities: private agri-food governance and the restructuring of the Kenyan horticulture industry', *Economic Geography* 86(2): 197–222.

Raikes, P., Jensen, M.F. and Ponte, S. (2000) 'Global commodity chain analysis and the French filiére approach: comparison and critique', *Economy and Society* 29(3): 390–417.

Riisgaard, L. (2009) 'Global Value Chains, labour organization and private social standards: lessons from East African cut flower industries', *World Development* 37(2): 326–340.

Schmitz, H. (1999) 'Global competition and local cooperation: success and failure in the Sinos Valley, Brazil', *World Development* 27(9): 1627–1650.

Sturgeon, T., Van Biesebroeck, J. and Gereffi, G. (2008) 'Value chains, networks and clusters: reframing the global automotive industry', *Journal of Economic Geography* 8(3): 297–321.

5 Socio-Cultural Contexts of Economic Change

5.1 Culture

Amin, A. and Thrift, N. (2004) *Cultural Economy Reader*. Malden, MA and Oxford: Blackwell.

Amin, A. and Thrift, N. (2007) 'Cultural-economy and cities', *Progress in Human Geography* 31(2): 143–161.

Aoyama, Y. (2007) 'The role of consumption and globalization in a cultural industry: the case of flamenco', *Geoforum* 38(1): 103–113.

Aoyama, Y. (2009) 'Artists, tourists, and the state: cultural tourism and the flamenco industry in Andalusia, Spain', *International Journal of Urban and Regional Research* 33(1) (March): 80–104.

Appadurai, A. (1996) *Modernity at Large: Cultural Dimensions of Globalization*. Minneapolis, MN: University of Minnesota Press.

Barnes, T.J. (2001) 'Retheorizing economic geography: from the quantitative revolution to the "cultural turn"', *Annals of the Association of American Geographers* 91(3) (September): 546–565.

Bourdieu, P. (1984) *Distinction: a Social Critique of the Judgment of Taste*. Cambridge, MA: Harvard University Press.

Butler, J. (1990) *Gender Trouble: Feminism and the Subversion of Identity*. New York: Routledge.

Castree, N. (2004) 'Economy and culture are dead! Long live economy and culture!', *Progress in Human Geography* 28(2): 204–226.

Chang, T.C. and B. Yeoh (1999) '"New Asia – Singapore": Communicating local cultures through global tourism', *Geoforum* 30(2): 101–115.

Christopherson, S. (2002) 'Why do labor market practices continue to diverge in a global economy? The "missing link" of investment rules', *Economic Geography* 78(1): 1–20.

Christopherson, S. (2004) 'The divergent worlds of new media: how policy shapes work in the creative economy', *Review of Policy Research* 21(4): 543–558.

Coe, N. (2001) 'A hybrid agglomeration? The development of a satellite-Marshallian industrial district in Vancouver's film industry', *Urban Studies* 38(10) (September): 1753–1775.

251

References

Connell, J. and Gibson, C. (2003) *Sound Tracks: Popular Music, Identity and Place*. London and New York: Routledge.

Crang, P. (1994) 'It's showtime: on the workplace geographies of display in a restaurant in southeast England', *Environment and Planning D: Society and Space* 12(6): 675–704.

Crewe, L. (2003) 'Markets in motion: geographies of retailing and consumption III', *Progress in Human Geography* 27(3): 352–362.

Dear, M. (1988) 'The postmodern challenge: reconstructing human geography', *Transactions, Institute of British Geographers* 13(3): 262–274.

Derrida, J. (1967) *Of Grammatology*. Baltimore, MD: Johns Hopkins University Press.

Dixon, D.P. and Jones, J.P. (1996) 'For a supercalifragilisticexpialidocious scientific geography', *Annals of the Association of American Geographers* 86(4): 767–779.

Ettlinger, N. (2003) 'Cultural economic geography and a relational and microspace approach to trusts, rationalities, networks, and change in collaborative workplaces', *Journal of Economic Geography* 3(2): 145–117.

Florida, R. (2002) 'The economic geography of talent', *Annals of the Association of American Geographers* 92(4): 743–755.

Foucault, M. (1981) 'The order of discourse', in R. Young (ed.) *Untying the Text: A Poststructuralist Reader*. Boston: Routledge. pp. 48–78.

Gertler, M.S. (2004) *Manufacturing Culture: The Institutional Geography of Industrial Practice*. Oxford: Oxford University Press.

Graham-Gibson, J.K. (2000) 'Poststructural interventions', in E. Sheppard and T.J. Barnes (eds) *A Companion to Economic Geography*. Oxford: Blackwell. pp. 95–110.

Gibson, C. and Kong, L. (2005) 'Cultural economy: a critical review', *Progress in Human Geography* 29(5): 541–561.

Gregson, N. and Crewe, L. (2003) *Second Hand Cultures*. Oxford: Berg.

Hall, P.A. and Soskice, D. (eds) (2001) *Varieties of Capitalism: The Institutional Foundations of Comparative Advantage*. Oxford: Oxford University Press.

Hall, P.G. (1998) *Cities in Civilisation*. London: Fromm.

Harvey, D. (1989) *The Condition of Postmodernity: An Enquiry into the Origins of Cultural Change*. Oxford: Basil Blackwell.

Izushi, H. and Aoyama, Y. (2006) 'Industry evolution and cross-sectoral skill transfers: a comparative analysis of the video game industry in Japan, the United States, and the United Kingdom', *Environment and Planning A* 38(10) (October): 1843–1861.

Jackson, P. (2002) 'Commercial cultures: transcending the cultural and the economic', *Progress in Human Geography* 26(1): 3–18.

Latour, B. (1987) *Science in Action: How to Follow Scientists and Engineers through Society*. Cambridge, MA: Harvard University Press.

Lawton Smith, H. (2003) 'Local innovation assemblages and institutional capacity in local high-tech economic development: the case of Oxfordshire', *Urban Studies* 40(7): 1353–1369.

Leslie, D. and Rantisi, N.M. (2006) 'Governing the design sector in Montréal', *Urban Affairs Review* 41: 309–337.

Leslie, D., and Reimer, S. (2003) 'Fashioning furniture: restructuring the furniture commodity chain', *Area* 35(4): 427–437.

Markusen, A. (1999) 'Fuzzy concepts, scanty evidence, policy distance: the case for rigor and policy relevance in critical regional studies', *Regional Studies* 33(9): 869–884.

References

Markusen, A. and Schrock, G. (2006) 'The artistic dividend: urban artistic specialization and economic development implications', *Urban Studies* 43(10): 1661–1686.

Marshall, A. (1920 [1890]) *Principles of Economics*. London: Macmillan (originally published in 1890).

Martin, R. (2001) 'Geography and public policy: the case of the missing agenda', *Progress in Human Geography* 25(2): 189–210.

Martin, R. and Sunley, P. (2001) 'Rethinking the "economic" in economic geography: broadening our vision or losing our focus?', *Antipode* 33(2): 148–161.

McDowell, L. and Court, G. (1994) 'Missing subjects: gender, power and sexuality in merchant banking', *Economic Geography* 70: 229–251.

Mitchell, D. (1995) 'There's no such thing as culture', *Transactions of the Institute of British Geographers* 20: 102–116.

Mitchell, K. (1995) 'Flexible circulation in the Pacific Rim: capitalisms in cultural context', *Economic Geography* 71(4): 364–382.

Molotch, H. (2002) 'Place in product', *International Journal of Urban and Regional Research* 26(4): 665–688.

North, D.C. (1990) *Institutions, Institutional Change and Economic Performance*. Cambridge: Cambridge University Press.

O'Neill, P.M. and Gibson-Graham, J.K. (1999) 'Enterprise discourse and executive talk: stories that destabilize the company', *Transactions, Institute of British Geographers* 24: 11–22.

Pollard, J. (2004) 'From industrial district to "urban village"? Manufacturing, money and consumption in Birmingham's jewellery quarter', *Urban Studies* 41(1): 173–194.

Power, D. (2002) '"Cultural industries" in Sweden: an assessment of their place in the Swedish economy', *Economic Geography* 78(2): 103–127.

Pratt, A. (1997) 'The cultural industries production system: a case study of employment change in Britain 1984–91', *Environment and Planning A* 29: 1953–1974.

Pratt, G. (1999) 'From registered nurse to registered nanny: discursive geographies of Filipina domestic workers in Vancouver, B.C.', *Economic Geography* 75(3): 215–236.

Rantisi, N.M. (2004) 'The ascendance of New York fashion', *International Journal of Urban and Regional Research* 28(1): 86–106.

Ray, L. and Sayer, A. (1999) *Culture and Economy After the Cultural Turn*. London: Sage Publications.

Rutherford, T.D. (2004) 'Convergence, the institutional turn and workplace regimes: the case of lean production', *Progress in Human Geography* 28(4): 425–466.

Saxenian, A. (1994) *Regional Advantage: Culture and Competition in Silicon Valley and Route 128*. Cambridge, MA: Harvard University Press.

Saxenian, A. (2006) *The New Argonauts: Regional Advantage in a Global Economy*. Cambridge, MA: Harvard University Press.

Sayer, A. (2003) '(De)commodification, consumer culture, and moral economy', *Environment and Planning D: Society and Space* 21: 341–357.

Schoenberger, E. (1997) *The Cultural Crisis of the Firm*. Cambridge, MA: Blackwell Publishers.

Scott, A.J. (1997) 'The cultural economy of cities', *International Journal of Urban and Regional Research* 21(2) (June): 323–339.

Scott, A.J. (2005) *On Hollywood: The Place, the Industry*. Princeton, NJ: Princeton University Press.

Soja, E. (1989) *Postmodern Geographies: The Reassertion of Space in Critical Social Theory*. London: Verso.

Storper, M. (2000) 'Conventions and institutions: rethinking problems of state reform, governance and policy', in L. Burlamaqui, A.C. Castro and H.-J. Chang (eds) *Institutions and the Role of the State*. Cheltenham: Edward Elgar. pp. 73–102.

Terry, W.C. (2009) 'Working on the water: on legal space and seafarer protection in the cruise industry', *Economic Geography* 85(4): 463–482.

Thrift, N. (2000a) 'Pandora's box? Cultural geographies of economies', in G. Clark, M. Feldman and M. Gertler (eds) *The Oxford Handbook of Economic Geography*. Oxford: Oxford University Press. pp. 689–704.

Thrift, N. (2000b) 'Performing cultures in the new economy', *Annals of the Association of American Geographers* 90(4): 674–692.

Thrift, N. and Olds, K. (1996) 'Reconfiguring the economic in economic geography', *Progress in Human Geography* 20: 311–337.

Urry, J. (1995) *Consuming Places*. London: Routledge.

Vinodrai, T. (2006) 'Reproducing Toronto's design ecology: career paths, intermediaries, and local labor markets', *Economic Geography* 82(3): 237–263.

Yapa, L. (1998) 'The poverty discourse and the poor in Sri Lanka', *Transactions, the Institute of British Geographers* 23: 95–115.

5.2 Gender

Bryceson, D.F. (2002) 'The scramble in Africa: reorienting rural livelihoods', *World Development* 30(5): 725–739.

Carney, J. (1993) 'Converting the wetlands, engendering the environment: the intersection of gender with agrarian change in the Gambia', *Economic Geography* 69: 329–348.

England, K. (1993) 'Suburban pink-collar ghettos: the spatial entrapment of women?', *Annals of the Association of American Geographers* 83: 225–242.

English, A. and Hegewisch, A. (2008) 'Still a man's labour market: the long-term earnings gap,' Washington, DC: Institute for Women's Policy Research.

Fairclough, Norman (1992) *Discourse and Social Change*. Cambridge: Polity Press.

Gibson-Graham, J.K. (1994) '"Stuffed if I know": reflections on post-modern feminist social research', *Gender, Place, and Culture* 1(2): 205–224.

Gray, M. and James, A. (2007) 'Connecting gender and economic competitiveness: lessons from Cambridge's high-tech regional economy', *Environment and Planning A* 39(2): 417–436.

Hanson, S. and Hanson, P. (1980) 'Gender and urban activity patterns in Uppsala, Sweden', *Geographical Review* 70: 291–299.

Hanson, S. and Pratt, G. (1988) 'Reconceptualizing the links between home and work in urban geography', *Economic Geography* 64: 299–321.

Hanson, S. and Pratt, G. (1992) 'Dynamic dependencies: a geographic investigation of local labour markets', *Economic Geography* 68: 373–405.

Hapke, H. and Ayyankeril, D. (2004) 'Gender, the work-life course, and livelihood strategies in a South Indian fish market', *Gender, Place, and Culture* 11: 229–256.

References

Jacobs, J. (1999) 'The sex segregation of occupations,' in G. Powell (ed.) *Handbook of Gender and Work*. Thousand Oaks, CA: Sage. pp. 125–141.

Jones, A. (1998) '(Re)producing gender cultures: theorizing gender in investment banking recruitment', *Geoforum* 29(4): 451–474.

Lawson, V. (2007) 'Geographies of care and responsibility', *Annals of the Association of American Geographers* 97(1): 1–11.

Mattingly, O.J. (2001) 'The home and the world domestic service and international networks of caring labor', *Annals of the Association of American Geographers* 91(2): 370–386.

McDowell, L. (1993) 'Space, place, and gender relations: part 2. identity, difference, feminist geometries and geographies', *Progress in Human Geography* 17(3): 305–318.

McDowell, L. (1997) *Capital Culture: Gender at Work in the City*. Oxford: Blackwell.

McDowell, L. (1999) *Gender, Identity, and Place: Understanding Feminist Geographies*. Cambridge: Polity Press.

McDowell, L. (2005) 'Men and the boys: bankers, burger makers, and barmen', in B. van Hoven and K. Horschelmann (eds) *Spaces of Masculinities*. New York: Routledge.

Meier, V. (1999) 'Cut-flower production in Colombia – a major development success story for women?', *Environment and Planning A* 31(2): 273–289.

Mullings, B. (2004) 'Globalization and the territorialization of the new Caribbean service economy', *Journal of Economic Geography* 4: 275–298.

Nagar, R. (2000) 'Mujhe jawab do! (Answer me!): women's grass-roots activism and social spaces in Chitrakoot (India)', *Gender, Place, and Culture* 7(4): 341–362.

Nelson, J. (1993) 'The study of choice or the study of provisioning? Gender and the definition of economies', in M. Ferber and J. Nelson (eds) *Beyond Economic Man: Feminist Theory and Economics*. Chicago: University of Chicago Press.

Nelson, K. (1986) 'Female labour supply characteristics and the suburbanization of low-wage office work', in M. Storper and A. Scott (eds) *Production, Work, and Territory*. Boston: Allen and Unwin.

Pavlovskaya, M. (2004) 'Other transitions: multiple economies of Moscow households in the 1990s', *Annals of the Association of American Geographers* 94(2): 329–351.

Pratt, G. (1999) 'From registered nurse to registered nanny: discursive geographies of Filipina domestic workers in Vancouver, B.C.', *Economic Geography* 75(3): 215–236.

Pratt, G. (2004) *Working Feminism*. Edinburgh: Edinburgh University Press.

Schroeder, R.A. (1999) *Shady Practices: Agroforestry and Gender Politics in The Gambia*. Berkeley, CA: University of California Press.

Scott, J. (1986) 'Gender: a useful category of historical analysis', *American Historical Review* 91: 1053–1075.

Seppala, P. (1998) *Diversification and Accumulation in Rural Tanzania: Anthropological Perspectives on Village Economics*. Uppsala, Sweden: Nordiska Afrikainstitutet.

Sheppard, E. (2006) 'The economic geography project', in S. Bagchi-Sen and H. Lawton Smith (eds) *Economic Geography: Past, Present, and Future*. New York: Routledge. pp. 11–23.

Silvey, R. and Elmhirst, R. (2003) 'Engendering social capital: women workers and rural-urban networks in Indonesia's crisis', *World Development* 31: 865–879.

Tivers, J. (1985) *Women Attached: The Daily Lives of Women with Young Children.* London: Croom Helm.

Wong, M. (2006) 'The gendered politics of remittances in Ghanaian transnational families', *Economic Geography* 82(4): 355–382.

Wright, M. (1997) 'Crossing the factory frontier: gender, place, and power in the Mexican *maquiladora*', *Antipode* 29(3): 278–302.

5.3 Institutions

Amin, A. (1999) 'An institutionalist perspective on regional economic development', *International Journal of Urban and Regional Research* 23(2): 365–378.

Amin, A. and Thrift, N. (1993) 'Globalization, institutional thickness, and local prospects', *Revue d'Economie Regionale et Urbaine* 3: 405–427.

Bathelt, H. (2003) 'Geographies of production: growth regimes in spatial perspective 1: innovation, institutions and social systems', *Progress in Human Geography* 27(6): 763–778.

Bathelt, H. (2006) 'Geographies of production: growth regimes in spatial perspective 3: toward a relational view of economic action and policy', *Progress in Human Geography* 30(2): 223–236.

Blake, M.K. and Hanson, S. (2005) 'Rethinking innovation: context and gender', *Environment and Planning A* 37(4): 681–701.

Boschma, R.A. and Frenken, K. (2006) 'Why is economic geography not an evolutionary science? Towards an evolutionary economic geography', *Journal of Economic Geography* 6(3): 273–302.

Essletzbichler, J. and Rigby, D.L. (2007) 'Exploring evolutionary economic geographies', *Journal of Economic Geography* 7: 549–571.

Florida, R. (1995) 'Toward the learning region', *Futures* 27(5): 527–536.

Florida, R. and Kenney, M. (1994) 'Institutions and economic transformation: the case of postwar Japanese capitalism', *Growth and Change* 25(2): 247–262.

Gertler, M.S. (2001) 'Best practice? Geography, learning and the institutional limits to strong convergence', *Journal of Economic Geography* 1(1): 5–26.

Gertler, M.S., Wolfe, D.A. and Garkut, D. (2000) 'No place like home? The embeddedness of innovation in a regional economy', *Review of International Political Economy* 7(4): 688–718.

Gray, M. and James, A. (2007) 'Connecting gender and economic competitiveness: lessons from Cambridge's high-tech regional economy', *Environment and Planning A* 39(2): 417–436.

Henry, N. and Pinch, S. (2001) 'Neo-Marshallian nodes, institutional thickness, and Britain's "Motor Sport Valley": thick or thin?', *Environment and Planning A* 33: 1169–1183.

Hudson, R. (2004) 'Conceptualizing economies and their geographies: spaces, flows and circuits', *Progress in Human Geography* 28(4): pp. 447–471.

Jessop, B. (2001) 'Institutional re(turns) and the strategic-relational approach', *Environment and Planning A* 33: 1213–1235.

Morgan, K. (1997) 'The learning region: institutions, innovation and regional renewal', *Regional Studies* 31: 491–503.

Munir, K.A. (2002) 'Being different: how normative and cognitive aspects of institutional environments influence technology transfer', *Human Relations* 55(12): 1403–1428.

References

Nelson, R.R. and Winter, S.G. (1982) *An Evolutionary Theory of Economic Change*. Cambridge, MA: Belknap Press of Harvard University Press.

North, D. (1990) *Institutional Change and Economic Performance*. Cambridge: Cambridge University Press.

Peet, J.R. (2007) *Geography of Power: The Making of Global Economic Policy*. London: Zed Press.

Powell, W.W. and Dimaggio, P.J. (eds) (1991) *The New Institutionalism in Organizational Analysis*. Chicago: University of Chicago Press.

Power, D. and Hauge, A. (2008) 'No man's brand: brands, institutions, and fashion', *Growth and Change* 39(1): 123–143.

Putnam, R.D. (1993) *Making Democracy Work: Civic Traditions in Modern Italy*. Princeton, NJ: Princeton University Press.

Rodríguez-Pose, A. and Storper, M. (2006) 'Better rules or stronger communities: on the social foundations of institutional change and its economic effects', *Economic Geography* 82(1): 1–25.

Scott, W.R. (1995) *Institutions and Organizations*. Thousand Oaks, CA: Sage Publications.

Storper, M. (1995) 'The resurgence of regional economics, ten years later: the region as a nexus of untraded interdependencies', *European Urban and Regional Studies* 2(3): 191–221.

Storper, M. (1997) *The Regional World: Territorial Development in a Global Economy*. New York: Guilford Press.

Sunley, P. (2008) Relational economic geography: a partial understanding or a new paradigm?', *Economic Geography* 84(1): 1–26.

Truffer, B. (2008) 'Society, technology, and region: contributions from the social study of technology to economic geography', *Environment and Planning A* 40: 966–985.

Veblen, T. (1915) 'The opportunity of Japan', *Journal of Race Development*, 6 (July): 23–38.

Veblen, T. (1925) 'Economic theory in the calculable future', *American Economic Review* 15(1S): 48–55.

Weber, M. (1905 [1998]) *The Protestant Work Ethic and the Spirit of Capitalism*. Los Angeles: Roxbury.

Weber, M. (1958 [1987]) *General Economic History*. New York: Greenburg.

Williamson, O. (1985) *The Economic Institutions of Capitalism: Firms, Markets, Relational Contracting*. New York: Free Press.

World Bank (2001) *World Bank Development Report: Building Institutions for Markets*. New York: Oxford University Press.

Yeung, H.W. (2000) 'Organizing "the firm" in industrial geography I: networks, institutions and regional development', *Progress in Human Geography* 24(2): 301–315.

Yeung, H.W. (2005) 'Rethinking relational economic geography', *Transactions of the Institute of British Geographers* 30(1): 37–51.

5.4 Embeddedness

Amin, A. and Cohendet, P. (2004) *Architectures of Knowledge: Firms, Capabilities, and Communities*. Oxford: Oxford University Press.

Amin, A. and Roberts, J. (2008) 'Knowing in action: beyond communities of practice', *Research Policy* 37(2): 353–369.

References

Bellandi, M. (2001) 'Local development and embedded large firms', *Entrepreneurship and Regional Development* 13: 189–210.

Burt, R.S. (1992) *Structural Holes*. Cambridge, MA: Harvard University Press.

Coe, N.M. (2000) 'The view from out West: embeddedness, inter-personal relations and the development of an indigenous film industry in Vancouver', *Geoforum* 31(4): 391–407.

Crouch, C. and Streeck, W. (1997) 'The future of capitalist diversity', in C. Rouch and W. Streeck (eds) *Political Economy of Modern Capitalism*. London: Sage. pp. 1–31.

DiMaggio, P. and Louch, H. (1998) 'Socially embedded consumer transactions: for what kinds of purchases do people most often use networks?', *American Sociological Review* 63(5): 619–637.

Eich-Born, M. and Hassink, R. (2005) 'On the battle between shipbuilding regions in Germany and South Korea', *Environment and Planning A* 37: 635–656.

Ettlinger, N. (2003) 'Cultural economic geography and a relational and microspace approach to trusts, rationalities, networks, and change in collaborative workplaces', *Journal of Economic Geography* 3(2): 145–171.

Evans, P.B. (1995) *Embedded Autonomy: States and Industrial Transformation*. Princeton, NJ: Princeton University Press.

Gertler, M.S. (2003) 'Tacit knowledge and the economic geography of context, or The undefinable tacitness of being (there)', *Journal of Economic Geography* 3(1): 75–99.

Gertler, M.S., Wolfe, D.A. and Garkut, D. (2000) 'No place like home? The embeddedness of innovation in a regional economy', *Review of International Political Economy* 7(4): 688–718.

Glückler, J. (2005) 'Making embeddedness work: social practice institutions in foreign consulting markets', *Environment and Planning A* 37(10): 1727–1750.

Grabher, G.E. (1993) 'The weakness of strong ties. The lock-in of regional development in the Ruhr area', in G. Grabher (ed.) *The Embedded Firm: On the Socioeconomics of Industrial Networks*. London: Routledge. pp. 255–277.

Granovetter, M.S. (1973) 'The strength of weak ties', *American Journal of Sociology* 78(6): 1360–1380.

Granovetter, M.S. (1985) 'Economic action and social structure: the problem of embeddedness', *American Journal of Sociology* 91(3): 481–510.

Hanson, S. and Blake, M. (2009) 'Gender and entrepreneurial networks', *Regional Studies* 43(1): 135–149.

Hanson, S. and Pratt, G. (1991) 'Job search and the occupational segregation of women', *Annals of the Association of American Geographers* 81: 229–253.

Harrison, B. (1992) 'Industrial districts: old wine in new bottles?', *Regional Studies* 26(5): 469–483.

Harrison, B. (1994) *Lean and Mean: The Changing Landscape of Corporate Power in the Age of Flexibility*. New York: Basic Books.

Hassink, R. (2007) 'The strength of weak lock-ins: the renewal of the Westmünsterland textile industry', *Environment and Planning A* 39: 1147–1165.

Hervas-Oliver, J.L. and Albors-Garrigos, J. (2009) 'The role of the firm's internal and relational capabilities in clusters: when distance and embeddedness are not enough to explain innovation', *Journal of Economic Geography* 9(2): 263–283.

References

Hess, M. (2004) '"Spatial" relationships? Towards a reconceptualization of embeddedness', *Progress in Human Geography* 28(2): 165–186.

Hughes, A., Wrigley, N. and Buttle, M. (2008) 'Global production networks, ethical campaigning, and the embeddedness of responsible governance', *Journal of Economic Geography* 8(3): 345–367.

Ingram, P., Robinson, J. and Busch, M.L. (2005) 'The intergovernmental network of world trade: IGO connectedness, governance, and embeddedness', *American Journal of Sociology* 111(3): 824–858.

Izushi, H. (1997) 'Conflict between two industrial networks: technological adaptation and inter-firm relationships in the ceramics industry in Seto, Japan', *Regional Studies* 31(2): 117–129.

James, A. (2007) 'Everyday effects, practices and causal mechanisms of "cultural embeddedness": learning from Utah's high tech regional economy', *Geoforum* 38(2): 393–413.

Jones, A. (2008) 'Beyond embeddedness: economic practices and the invisible dimensions of transnational business activity', *Progress in Human Geography* 32(1): 71–88.

Krippner, G.R. and Alvarez, A.S. (2007) 'Embeddedness and the intellectual projects of economic sociology', *Annual Review of Sociology* 33(1): 219–240.

Kus, B. (2006) 'Neoliberalism, institutional change and the welfare state: the case of Britain and France', *International Journal of Comparative Sociology* 47(6): 488–525.

Liu, W. and Dicken, P. (2006) 'Transnational corporations and "obligated embeddedness": Foreign direct investment in China's automobile industry', *Environment and Planning A* 38(7): 1229–1247.

Lowe, N. (2009) 'Challenging tradition: unlocking new paths to regional industrial upgrading', *Environment and Planning A* 41(1): 128–145.

MacKinnon, D., Cumbers, A. and Chapman, K. (2002) 'Learning, innovation and regional development: a critical appraisal of recent debates', *Progress in Human Geography* 26(3): 293–311.

Morgan, K. (2005) 'The exaggerated death of geography: learning, proximity and territorial innovation systems', *Journal of Economic Geography* 4(1): 3–21.

Oinas, P. (1999) 'Voices and silences: the problem of access to embeddedness', *Geoforum* 30(4): 351–361.

Parsons, T. and Smelser, N. (1956) *Economy and Society: A Study in the Integration of Economic and Social Theory.* Glencoe, IL: The Free Press.

Peck, J. (2005) 'Economic sociologies in space', *Economic Geography* 81(2): 129–175.

Polanyi, K. (1944) *The Great Transformation.* Boston: Beacon Press.

Polanyi, M. (1967) *The Tacit Dimension.* New York: Anchor Books.

Portes, A. and Sensenbrenner, J. (1993) 'Embeddedness and immigration: notes on the social determinants of economic action', *American Journal of Sociology* 98(6): 1320–1350.

Uzzi, B. (1996) 'The sources and consequences of embeddedness for the economic performance of organizations: the network effect', *American Sociological Review* 61(4): 674–698.

Wenger, E. (1998) *Communities of Practice: Learning, Meaning and Identity.* Cambridge: Cambridge University Press.

Williamson, O. (1985) *The Economic Institutions of Capitalism: Firms, Markets, Relational Contracting.* New York: Free Press.

5.5 Networks

Amin, A. (1999) 'An institutionalist perspective on regional economic development', *International Journal of Urban and Regional Research* 23(2): 365–378.

Amin, A. and Thrift, N. (1992) 'Neo-Marshallian nodes in global networks', *International Journal of Urban and Regional Research* 16: 571–587.

Anderson, A.R. and Jack, S. (2002) 'The articulation of social capital in entrepreneurial networks: a glue or lubricant?', *Entrepreneurship and Regional Development* 14: 193–210.

Bathelt, H. (2006) 'Geographies of production: growth regimes in spatial perspective 3: Toward a relational view of economic action and policy', *Progress in Human Geography* 30(2): 223–236.

Bathelt, H. and Glückler, J. (2003) 'Toward a relational economic geography', *Journal of Economic Geography* 3(2): 117–144.

Beaverstock, J.V., Smith, R.G. and Taylor, P.J. (1999) 'A roster of world cities', *Cities* 16(6): 445–458.

Bebbington, A. (2003) 'Global networks and local developments: agendas for development geography', *Tijdschrift voor Economische en Sociale Geografie* 94(3): 297–309.

Bek, D., McEwan, C. and Bek, K.E. (2007) 'Ethical trading and socioeconomic transformation: Critical reflections on the South African wine industry', *Environment and Planning A* 39(2): 301–319.

Berndt, C. and Boeckler, M. (2009) 'Geographies of circulation and exchange: constructions of markets', *Progress in Human Geography* 33(4): 535–551.

Callon, M. (1986) 'Some elements of a sociology of translation: domestication of the scallops and the fishermen of St Brieuc Bay', in J. Law (ed.) *Power, Action, and Belief: A New Sociology of Knowledge.* London: Routledge & Kegan Paul. pp. 196–233.

Callon, M. (1999) 'Actor-network theory-the market test', in J. Law and J. Hassard (eds) *Actor Network Theory and After.* Oxford: Blackwell. pp. 181–195.

Callon, M. and Muniesa, F. (2005) 'Peripheral vision: economic markets as calculative collective devices', *Organization Studies* 26(8): 1229–1250.

Camagni, R. (ed.) (1991) *Innovation Networks: Spatial Perspectives.* London: Belhaven.

Castells, M. (1996) *The Rise of the Network Society.* Oxford: Blackwell.

Coe, N.M., Hess, M. Yeung, H.W., Dicken, P. and Henderson, J. (2004) '"Globalizing" regional development: a global production networks perspective', *Transactions of the Institute of British Geographers* 29(4): 468–484.

Cooke, P. and Morgan, K. (1998) *The Associational Economy: Firms, Regions, and Innovation.* Oxford: Oxford University Press.

Dicken, P., Kelly, P.F., Olds, K. and Yeung, H.W. (2001) 'Chains and networks, territories and scales: towards a relational framework for analyzing the global economy', *Global Networks* 1(2): 89–112.

Friedmann, J. (1986) 'The world city hypothesis', *Development and Change* 17(1): 69–84.

Glückler, J. (2007) 'Economic geography and the evolution of networks', *Journal of Economic Geography* 7(5): 619–634.

Grabher, G. (2006) 'Trading routes, bypasses, and risky intersections: mapping the travels of "networks" between economic sociology and economic geography', *Progress in Human Geography* 30(2): 163–189.

References

Granovetter, M. (1985) 'Economic action and social structure: the problem of embeddedness', *American Journal of Sociology* 91(3): 481–510.

Hadjimichalis, C. and Hudson, R. (2006) 'Networks, regional development and democratic control', *International Journal of Urban and Regional Research* 30(4): 858–872.

Hanson, S. (2000) 'Networking', *Professional Geographer* 52(4): 751–758.

Hanson, S. and Blake, M. (2009) 'Gender and entrepreneurial networks', *Regional Studies* 43(1): 135–149.

Harrison, B. (1992) 'Industrial districts: old wine in new bottles?', *Regional Studies* 26(5): 469–483.

Henry, L., Mohan, G. and Yanacopulos, H. (2004) 'Networks as transnational agents of development', *Third World Quarterly* 25(5): 839–855.

Hess, M. (2004) '"Spatial" relationships? Towards a reconceptualization of embeddedness', *Progress in Human Geography* 28(2): 165–186.

Kim, S.J. (2006) 'Networks, scale, and transnational corporations: the case of the South Korean seed industry', *Economic Geography* 82(3): 317–338.

Kingsley, G. and Malecki, E.J. (2004) 'Networking for competitiveness', *Small Business Economics* 23(1): 71–84.

Knox, P.L. and Taylor, P.J. (eds) (1995) *World Cities in a World-system.* Cambridge: Cambridge University Press.

Knorr Cetina, K. and Bruegger, U. (2002) 'Global microstructures: the virtual societies of financial markets', *American Journal of Sociology* 107(4): 905–950.

Kuhn, T. (1962) *The Structure of Scientific Revolutions.* Chicago: Chicago University Press.

Latour, B. (1991) 'Technology is society made durable', in J. Law (ed.) *A Sociology of Monsters: Essays on Power, Technology, and Domination.* London: Routledge. pp. 103–131.

Law, J. (1992) 'Notes on the theory of the actor network: ordering, strategy, and heterogeneity', *Systems Practice* 5(4): 379–393.

Law, J. (2008) 'Actor-network theory and material semiotics', in B.S. Turner (ed.) *The New Blackwell Companion to Social Theory*, 3rd edn. Oxford: Blackwell. pp. 141–158.

Mackinnon, D., Chapman, K. and Cumbers, A. (2004) 'Networking, trust and embeddedness amongst SMEs in the Aberdeen oil complex', *Entrepreneurship and Regional Development* 16(2): 87–106.

Malecki, E.J. (2000) 'Soft variables in regional science', *Review of Regional Studies* 30(1): 61–69.

Malecki, E.J. and Tootle, D.M. (1997) 'Networks of small manufacturers in the USA: Creating embeddedness', in M. Taylor and S. Conti (eds) *Interdependent and Uneven Development: Global–Local Perspectives.* Aldershot: Ashgate. pp. 195–221.

Murdoch, J. (1995) 'Actor-networks and the evolution of economic forms: combining description and explanation in theories of regulation, flexible specialization, and networks', *Environment and Planning A* 27(5): 731–757.

Murdoch, J. (1998) 'The spaces of actor-network theory', *Geoforum* 29(4): 357–374.

Murphy, J.T. (2002) 'Networks, trust, and innovation in Tanzania's manufacturing sector', *World Development* 30(4): 591–619.

Murphy, J.T. (2006) 'Building trust in economic space', *Progress in Human Geography* 30(4): 427–450.

Nijkamp, P. (2003) 'Entrepreneurship in a modern network economy', *Regional Studies* 37(4): 395–405.

Peck, J. (2005) 'Economic sociologies in space', *Economic Geography* 81(2): 129–175.

Powell, W.W. (1990) 'Neither market or hierarchy: network forms of organization', *Research in Organizational Behavior* 12: 295–336.

Powell, W.W. and Smith-Doerr, L. (1994) 'Networks and economic life', in N.J. Smelser and R. Swedberg (eds) *The Handbook of Economic Sociology*, 1st edn. Princeton, NJ: Princeton University Press. pp. 368–402.

Powell, W.W. and Smith-Doerr, L. (2005) 'Networks and economic life', in N.J. Smelser and R. Swedberg (eds) *The Handbook of Economic Sociology*, 2nd edn. Princeton, NJ: Princeton University Press. (1st edn, 1994.) pp. 379–402.

Rossi, E.C. and Taylor, P.J. (2006) '"Gateway cities" in economic globalisation: how banks are using Brazilian cities', *Tijdschrift voor Economische en Sociale Geografie* 97(5): pp. 515–534.

Sheppard, E. (2002) 'The spaces and times of globalization: place, scale, networks, and positionality', *Economic Geography* 78(3): 307–330.

Smith, A. (2003) 'Power relations, industrial clusters, and regional transformations: pan-European integration and outward processing in the Slovak clothing industry', *Economic Geography* 79(1): 17–40.

Staber, U. (2001) 'The structure of networks in industrial districts', *International Journal of Urban and Regional Research* 25(3): 537–552.

Sunley, P. (2008) 'Relational economic geography: a partial understanding or a new paradigm?', *Economic Geography* 84(1): 1–26.

Taylor, P.J. and Aranya, R. (2008) 'A global "urban roller coaster"? Connectivity changes in the world city network 2000–2004', *Regional Studies* 42(1): 1–16.

Taylor, P.J., Derudder, B., Garcia, C.G. and Witlox, F. (2009) 'From North–South to "Global" South? An investigation of a changing "South" using airline flows between cities, 1970–2005', *Geography Compass* 3(2): 836–855.

Truffer, B. (2008) 'Society, technology, and region: contributions from the social study of technology to economic geography', *Environment and Planning A* 40(4): 966–985.

Turner, S. (2007) 'Small-scale enterprise livelihoods and social capital in Eastern Indonesia: ethnic embeddedness and exclusion', *Professional Geographer* 59(4): 407–420.

Woolcock, M. (1998) 'Social capital and economic development: toward a theoretical synthesis and policy framework', *Theory and Society* 27(2): 151–208.

Yeung, H.W.C. (2005) 'The firm as social networks: an organisational perspective', *Growth and Change* 36(3): 307–328.

6 Emerging Themes in Economic Geography

6.1 Knowledge Economy

Aoyama, Y. and Castells, M. (2002) 'An empirical assessment of the informational society: Employment and occupational structures of G-7 countries', *International Labour Review* 141: 123–159.

References

Aoyama, Y. and Sheppard, E. (2003) 'The dialectics of geographic and virtual spaces', *Environment and Planning A* 35(7): 1151–1156.

Asheim, B.T. and Coenen, L. (2005) 'Knowledge bases and regional innovation systems: comparing Nordic clusters', *Research Policy* 34: 1173–1190.

Aydalot, P. (1985) *High Technology Industry and Innovative Environments.* London: Routledge.

Baudrillard, J. (1970) *The Consumer Society.* London: Sage Publications.

Bell, D. (1973) *The Coming of Postindustrial Society: A Venture on Social Forecasting.* New York: Basic Books.

Beyers, W.B. (1993) 'Producer services', *Progress in Human Geography* 17(2): 221–231.

Beyers, W. (2002) 'Services and the new economy: elements of a research agenda', *Journal of Economic Geography* 2: 1–29.

Castells, M. (1989) *The Informational City: Information Technology, Economic Restructuring and the Urban-Regional Process.* London: Blackwell.

Castells, M. (2000) *The Rise of the Network Society*, 2nd edn. Oxford: Blackwell.

Castells, M. and Hall, P. (1994) *Technopoles of the World: The Making of Twenty-first-century Industrial Complexes.* London: Routledge.

Castells, M. and Aoyama, Y. (1994) 'Paths toward the informational society: employment structure in G-7 Countries, 1920–90', *International Labour Review* 133: 5–33.

Clark, G.L. (2002) 'London in the European financial services industry: locational advantage and product complementarities', *Journal of Economic Geography* 2(4): 433–453.

Cohen, S. and Zysman, J. (1987) *Manufacturing Matters. The Myth of the Postindustrial Economy.* New York: Basic Books.

Cooke, P. (2001) 'Regional innovation systems, clusters, and the knowledge economy', *Industrial and Corporate Change* 10(4): 945–974.

Cooke, P. (2007) *Growth Cultures: The Global Bioeconomy and Its Bioregions.* Abingdon: Routledge.

Daniels, P. (1985) *Service Industries.* London: Routledge.

Daniels, P. and Bryson, J. (2002) 'Manufacturing services and servicing manufacturing: changing forms of production in advanced capitalist economies', *Urban Studies* 39: 977–991.

Daniels, P.W. and Moulaert, F. (1991) *The Changing Geography of Advanced Producer Services: Theoretical and Empirical Perspectives.* London: Belhaven Press.

Drucker, P. (1969) *The Age of Discontinuity: Guidelines to Our Changing Society.* New York: Harper and Row.

Florida, R. (1995) 'Toward the learning region', *Futures* 27(5): 527–536.

Florida, R. (2002) 'The economic geography of talent. *Annals of the Association of American Geographers* 92(4): 743–755.

Florida, R. and Kenney, M. (1988) 'Venture capital, high technology and regional development', *Regional Studies* 22(1): 33–48.

Fuchs, V. (1968) *The Services Economy.* New York: Columbia University Press.

Gershuny, J.I. and Miles, I.D. (1983) *New Service Economy: The Transformation of Employment in Industrial Societies.* London: F. Pinter.

Gertler, M.S. and Levitte, Y.M. (2005) 'Local nodes in global networks: the geography of knowledge flows in biotechnology innovation', *Industry & Innovation* 12(4): 487–507.

263

References

Gillespie, A. and Green, A. (1987) 'The changing geography of producer services employment in Britain', *Regional Studies* 21(5): 397–411.

Glaeser, E. (2005) 'Review of Richard Florida's *The Rise of the Creative Class*', *Regional Science and Urban Economics* 35: 593–596.

Gordon, R.J. (2000) 'Does the "new economy" measure up to the great inventions of the past?', NBER Working Paper 7833. Cambridge, MA: National Bureau of Economic Research, August. Available at: http://www.nber.org/papers/w7833.

Hall, S. (2009) 'Ecologies of business education and the geographies of knowledge', *Progress in Human Geography* 33(5): 599–618.

Jones, A. (2005) 'Truly global corporations? Theorizing "organizational globalization" in advanced business-services', *Journal of Economic Geography* 5(2): 177–200.

Kuznets, S. (1971) *Economic Growth of Nations: Total Output and Production Structure*. Harvard, MA: Belknap.

Leamer, E.E. and Storper, M. (2001) 'The economic geography of the Internet age', *Journal of International Business Studies* 32.

Leslie, D. (1995) 'Global scan: the globalization of advertising agencies, concepts and campaigns', *Economic Geography* 71(4): 402–426.

Lorenzen, M. and Andersen, K.V. (2009) 'Centrality and creativity: does Richard Florida's creative class offer new insights into urban hierarchy?', *Economic Geography* 85: 363–390.

Lundquist, K., Olander, L.-O. and Henning, M. (2008) 'Producer services: growth and roles in long-term economic development', *The Service Industries Journal* 28: 463–477.

Machlup, F. (1962) *The Production and Distribution of Knowledge in the United States*. Princeton, NJ: Princeton University Press.

Malecki, E. (1997) *Technology and Economic Development: The Dynamics of Local, Regional and National Competitiveness*. Reading, MA: Addison-Wesley.

Marcuse, P. (2003) 'Review of *The Rise of the Creative Class* by Richard Florida', *Urban Land* 62: 40–1.

Marshall, J.N., Damesick, P., Wood, P. (1987) 'Understanding the location and role of producer services in the United Kingdom', *Environment and Planning A* 19(5): 575–595.

McQuaid, R.W. (2002) 'Entrepreneurship and ICT Industries: Support from regional and local policies', *Regional Studies* 36: 909–919.

Miles, I. (2000) 'Services innovation: Coming of age in the knowledge-based economy', *International Journal of Innovation Management* 4: 371–389.

Nelson, K. (1986) 'Labor demand, labor supply and the suburbanization of low-wage Office Work', in A.J. Scott and M. Storper (eds) *Production, Work and Territory*. Boston: Allen & Unwin, pp. 149–171.

Pandit, K. (1990a) 'Service labour allocation during development: longitudinal perspectives on cross-sectional patterns', *Annals of Regional Science* 24(1) (March): 29–41.

Pandit, K. (1990b) 'Tertiary sector hypertrophy during development: an examination of regional variation', *Environment and Planning A* 22(10): 1389–1405.

Peck, J. (2005) 'Struggling with the creative class', *International Journal of Urban and Regional Research* 29(4): 740–770.

Romer, P.M. (1986) 'Increasing returns and long-run growth', *Journal of Political Economy* 94(5): 1002–1037.

Sassen, S. (1991) *The Global City: New York, London, Tokyo*. Princeton, NJ:

Sayer, A. and Walker, R.A. (1993) *The New Social Economy: Reworking the Division of Labor*. Blackwell: Princeton University Press.

Saxenian, A. (1994) *Regional Advantage: Culture and Competition in Silicon Valley and Route 128*. Cambridge, MA: Harvard University Press.

Saxenian, A. (2006) *The New Argonauts: Regional Advantage in a Global Economy*. Cambridge, MA: Harvard University Press.

Saxenian, A. (2007) 'Brain circulation and regional innovation: Silicon Valley-Hsinchu-Shanghai Triangle', in K. Polenski (ed.) *The Economic Geography of Innovation*. Cambridge: Cambridge University Press. pp. 196–212.

Singelmann, J. (1977) *The Transformation of Industry: From Agriculture to Service Employment*. Beverly Hills, CA: Sage.

Stanback, T.M. (1980) *Understanding the Service Economy*. Baltimore, MD: Johns Hopkins University Press.

Stanback, T.M., Bearse, P.J., Noyelle, T. and Karasek, R.A. (1981) *Services: The New Economy*. Totowa, NJ: Allanheld, Osmun.

Storper, M. and Scott, A.J. (2009) 'Rethinking human capital, creativity and urban growth', *Journal of Economic Geography* 9(2): 147–167.

Thrift, N. (1997) 'The rise of soft capitalism', *Cultural Values* 1(1): 29–57.

Thrift, N. (1998) 'Virtual capitalism: the globalization of reflexive business knowledge', in J.G. Carrier and D. Miller (eds) *Virtualism: A New Political Economy*. Oxford, New York: Berg. pp.161–186.

Thrift, N. (2000) 'Performing cultures in the new economy', *Annals of the Association of American Geographers* 90(4): 674–692.

Touraine, A. (1969) *La Société Post-industrielle*. Paris: Denoël.

Warf, B. (1989) 'Telecommunications and the globalization of financial services', *Professional Geographer* 31: 257–271.

Wood, P. (2005) 'A service-informed approach to regional innovation – or adaptation?', *The Service Industries Journal* 25(4): 429–445.

Zook, M.A. (2001) 'Old hierarchies or new networks of centrality? The global geography of the internet content market', *American Behavioral Scientist* 44(10) (June): 1679–1696.

Zook, M. and Graham, M. (2007) 'The creative reconstruction of the internet: Google and the privatization of cyberspace and DigiPlace', *Geoforum* 38: 1322–1343.

6.2 Financialization

Agnes, P. (2000) 'The "end of geography" in financial services? Local embeddedness and territorialization in the interest rate swaps industry', *Economic Geography* 76(4): 347–366.

Ashton, P. (2009) 'An appetite for yield: the anatomy of the subprime mortgage crisis', *Environment and Planning A* 41(6): 1420–1441.

Boyer, R. (2000) 'Is finance-led growth regime a viable alternative to Fordism? A preliminary analysis', *Economy and Society* 29(1): 111–145.

Clark, G.L. (1998) 'Why convention dominates pension fund trustee investment decision-making', *Environment and Planning A* 30(6): 997–1015.

Clark, G.L. (2000) *Pension Fund Capitalism*. Oxford: Oxford University Press.

Clark, G.L. and Wójcik, D. (2005) 'Path dependence and financial markets: the economic geography of the German model, 1997–2003', *Environment and Planning A* 37: 1769–1791.

References

Clark, G.L. Mansfield, D. and Tickell, A. (2001) 'Emergent frameworks in global finance: accounting standards and German supplementary pensions', *Economic Geography* 77: 250–271.

Dicken, P. (1998) *Global Shift: Transforming the World Economy*, 3rd edn. London: Paul Chapman.

Dore, R. (2008) 'Financialization of the global economy', *Industrial and Corporate Change* 17(6): 1097–1112.

Dow, S.C. and Rodriguez-Fuentes, C.J. (1997) 'Regional finance: a survey', *Regional Studies* 31(9): 903–920.

Engelen, E. (2003) 'The logic of funding European pension restructuring and the dangers of financialisation', *Environment and Planning A* 35: 1357–1372.

Engelen, E. (2007) '"Amsterdamned"? The uncertain future of a financial centre', *Environment and Planning A* 39(6): 1306–1324.

Erturk, I., Froud, J., Sukhdev, J., Leaver, A. and Williams, K. (2007) 'The democratization of finance? Promises, outcomes and conditions', *Review of International Political Economy* 14(4): 553–575.

Fackler, M. (2007) 'Japanese housewives sweat in secret as markets reel', *New York Times*, 16 September. Accessed on 2 December 2008 at: http://www.nytimes.com/2007/09/16/business/worldbusiness/16housewives.html?em.

Florida, R. and Kenney, M. (1988) 'Venture capital, high technology and regional development', *Regional Studies* 22(1): 33–48.

Friedman, J. (1986) 'The world city hypothesis development and change', *Development and Change* 17(1): 69–84.

Grote, M.H., Lo, V. and Harrschar-Ehrnborg, S. (2002) 'A value chain approach to financial centers: the case of Frankfurt', *Tijdschrift voor Economische en Sociale Geografie* 93(4): 412–423.

Kindleberger, C.P. (1974) *The Formation of Financial Centres: A Study in Comparative Economic History*. Princeton, NJ: Princeton University Press.

Langley, P. (2006) 'The making of investor subjects in Anglo-American pensions', *Environment and Planning D: Society and Space* 24(6): 919–934.

Laulajainen, R. (2003) *Financial Geography: A Banker's View*. London: Routledge.

Lee, R. (1996) 'Moral money? LETS and the social construction of local economic geographies in Southeast England', *Environment and Planning A* 28: 1377–1394.

Lee, R. and Schmidt-Marwede, U. (1993) 'Interurban competition? Financial centres and the geography of financial production', *International Journal of Urban and Regional Research* 17: 492–515.

Leinbach, T., and Amrhein, C. (1987) 'A geography of the venture capital industry in the US', *The Professional Geographer* 39: 146–158.

Leyshon, A. and Thrift, N. (1997) *Money Space: Geographies of Monetary Transformation*. London: Routledge.

Markowitz, H.M. (1959) *Portfolio Selection: Efficient Diversification of Investments*. New York: John Wiley & Sons.

Marx, K. (1894) *Das Kapital: Kritik der politischen Oekonomie. Buch III: Der Gesamtprocess der Kapitalistischen Lrodvktion*. Hamburg: Verlag von Otto Meissner.

McLaughlin, A. (2008) 'Japanese housewife online traders', *Japan Inc. Magazine* 75, 15 January. Accessed on 2 December 2008 at: http://www.japaninc.com/mgz_jan-feb_2008_housewife-online-trading.

266

Nelson, K. (1986) 'Labor demand, labor supply and the suburbanization of low-wage office work', in A.J. Scott and M. Storper (eds) *Production, Work and Territory*. Boston: Allen & Unwin, pp. 149–171.

O'Brien, R. (1992) *Global Financial Integration: The End of Geography*. London: Pinter.

Pollard, J. and Samers, M. (2007) 'Islamic banking and finance: postcolonial political economy and the decentring of economic geography', *Transactions of the Institute of British Geographers* 32(3): 313–330.

Pryke, M. (1991) 'An international city going "global": spatial change in the City of London', *Environment and Planning D: Society and Space* 9: 197–222.

Pryke, M. (1994) 'Looking back on the space of a boom: redeveloping spatial matrices in the City of London', *Environment and Planning A* 26: 235–264.

Rimmer, P.J. (1986) 'Japan's world cities: Tokyo, Osaka, Nagoya or Tokaido megalopolis', *Development and Change* 17: 121–157.

Roberts, S.M. (1995) 'Small place, big money: the Cayman Islands and the international financial system', *Economic Geography* 71(3): 237–256.

Sassen, S. (1991) *The Global City: New York, London, Tokyo*. Princeton, NJ: Princeton University Press.

Strange, S. (1986) *Casino Capitalism*. Oxford: Basil Blackwell.

Tickell, A. (2000) 'Finance and localities', in G.L. Clark, M.P. Feldman and M.S. Gertler (eds) *The Oxford Handbook of Economic Geography*. Oxford: Oxford University Press: pp. 230–247.

Thrift, N. (1994) 'On the social and cultural determinants of international financial centres: the case of the City of London', in S. Corbridge, R. Martin and N. Thrift (eds) *Money, Power and Space*. Oxford and Cambridge, MA: Basil Blackwell. pp. 327–355.

Thrift, N. (2000) 'Less mystery, more imagination: the future of the City of London', *Environment and Planning A* 32: 381–384.

Warf, B. and Cox, J.C. (1995) 'U.S. Bank Failures and Regional Economic Structure', *The Professional Geographer* 47(1): 3–16.

Zook, M.A. (2004) 'The knowledge brokers: venture capitalists, tacit knowledge and regional development', *International Journal of Urban and Regional Research* (September): 621–641.

267

6.3 Consumption

Berry, B. (1967) *Geography of Market Centres and Retail Distribution*. Englewood Cliffs, NJ: Prentice-Hall.

Bhachu, P. (2004) *Dangerous Designs: Asian Women Fashion the Diaspora Economies*. London: Routledge.

Christaller, W. (1966) *Central Places in Southern Germany*. An English translation of *Die zentralen Orte in Süddeutschland* by C.W. Baskin, Englewood Cliffs, NJ: Prentice Hall (originally published in 1933).

Clarke, I., Hallsworth, A., Jackson, P., de Kervenoael, R., del Aguila, R.P. and Kirkup, M. (2006) 'Retail restructuring and consumer choice 1. Long-term local changes in consumer behaviour: Portsmouth, 1980–2002', *Environment and Planning A* 38(1): 25–46.

References

Coe, N.M. and Wrigley, N. (2007) 'Host economy impacts of transnational retail: the research agenda', *Journal of Economic Geography* 7(4): 341–371.

Connell, J. and Gibson, C. (2004) 'World Music: Deterritorialising place and identity', *Progress in Human Geography* 28(3): 342–361.

Crang, P. (1994) 'It's showtime: on the workplace geographies of display in a restaurant in southeast England', *Environment and Planning D: Society and Space* 12(6): 675–704.

Domosh, M. (1996) 'The feminized retail landscape: gender ideology and consumer culture in nineteenth-century New York City', in N. Wrigley and M. Lowe (eds) *Retailing, Consumption and Capital*. Harlow: Longman. pp. 257–70.

Gellately, R. (1974) *The Politics of Economic Despair: Shopkeepers and German Politics 1890–1914*. London: Sage.

Gerth, H.H. and Mills, C.W. (eds) (1946) *From Max Weber: Essays in Sociology*. New York: Oxford University Press.

Goldman, A. (1991) 'Japan distribution system: institutional structure, internal political economy, and modernization', *Journal of Retailing* 67(2): 154–184.

Goldman, A. (2001) 'The transfer of retail formats into developing economies: the example of China', *Journal of Retailing* 77(2): 221–242.

Goss, J. (2004) 'Geography of Consumption I', *Progress in Human Geography* 28(3): 369–380.

Goss, J. (2006) 'Geographies of consumption: the work of consumption', *Progress in Human Geography* 30(2): 237–249.

Gibson-Graham, J.K. (1996) *The End of Capitalism (As We Knew It): A Feminist Critique of Political Economy*. Oxford: Blackwell.

Grabher, G., Ibert, O. and Floher, S. (2008) 'The neglected king: the customer in the new knowledge ecology of innovation', *Economic Geography* 84(3): 253–280.

Gregson, N. and L. Crewe (1997) 'The bargain, the knowledge, and the spectacle: making sense of consumption in the space of the car-boot sale', *Environment and Planning D: Society and Space* 15(1): 87–112.

Gregson, N. and Crewe, L. (2003) *Second-hand Cultures*. New York: Berg.

Hotelling, H. (1929) 'Stability in competition', *The Economic Journal* 39(153): 41–57.

Hughes, A. and Reimer, S. (2004) *Geographies of Commodity Chains*. London: Routledge.

Hughes, A., Buttle, M. and Wrigley, N. (2007) 'Organisational geographies of corporate responsibility: a UK–US comparison of retailers' ethical trading initiatives', *Journal of Economic Geography* 7(4): 491–513.

Jackson, P. (2004) 'Local consumption cultures in a globalizing world', *Transactions of the Institute of British Geographers* 29(2): 169–178.

Kozul-Wright, Z. and Stanbury L. (1998) 'Becoming a globally competitive player: the case of the music industry in Jamaica', *UNCTAD Discussion Papers*, No. 138, October.

Larner, W. (1997) 'The legacy of the social: market governance and the consumer', *Economy and Society*, 26(3): 373–399.

Leslie, D. and Reimer, S. (2003) 'Fashioning Furniture: restructuring in the furniture commodity chain', *Area* 35(4): 427–437.

Leyshon, A. (2004) 'The limits to capital and geographies of money', *Antipode* 36(3): 461–469.

Marsden, T. and Wrigley, N. (1995) 'Regulation, retailing, consumption', *Environment and Planning A* 27(12): 1899–1912.

Marsden, T. and Wrigley, N. (1996) 'Retailing, the food system and the regulatory state', in N. Wrigley and M.S. Lowe (eds) *Retailing, Consumption and Capital*. Harlow, Essex: Longman. pp. 33–47.

Pine, B. and Gilmore, J. (1999) *The Experience Economy: Work is Theatre and Every Business a Stage*. Boston, MA: Harvard Business School Press.

Rostow, W.W. (1953) *The Process of Economic Growth*. Oxford: Oxford University Press.

Sayer, A. (2003) '(De)commodification, consumer culture, and moral economy', *Environment and Planning D: Society and Space* 21(3): 341–357.

Veblen, T. (1899) *The Theory of the Leisure Class: An Economic Study of Institutions*. New York: The Modern Library.

von Hippel, E. (2001) 'Innovation by user communities: learning from open-sources software', *MIT Sloan Management Review* 42(4): 82–86.

Wang, L. and Lo, L. (2007) 'Immigrant grocery shopping behaviour: ethnic identity versus accessibility', *Environment and Planning A* 39(3): 684–699.

Williams, C.C. (2002) 'Social exclusion in a consumer society: a study of five rural communities', *Social Policy and Society* 1(3): 203–211.

Wrigley, N., Lowe, M. and Currah, A. (2002) 'Retailing and e-tailing', *Urban Geography* 23(2): 180–197.

Zukin, S. (1995) *The Cultures of Cities*. Oxford: Blackwell.

6.4 Sustainable Development

Adams, W.M. (2009) *Green Development: Environment and Sustainability in a Developing World*, 3rd edn. London: Routledge.

Adger, W.N. (2003) 'Social capital, collective action, and adaptation to climate change', *Economic Geography* 79(4): 387–404.

Agyeman, J. and Evans, B. (2004) '"Just sustainability": the emerging discourse of environmental justice in Britain?', *Geographical Journal* 170(2): 155–164.

Angel, D.P. and Rock, M.T. (2003) 'Engaging economic development agencies in environmental protection: the case for embedded autonomy', *Local Environment* 8(1): 45–59.

Barr, S. and Gilg, A. (2006) 'Sustainable lifestyles: framing environmental action in and around the home', *Geoforum* 37(6): 906–920.

Basel Action Network (2009) Basel Action Network webpage: http://www.ban.org.

Bebbington, A. and Perreault, T. (1999) 'Social capital, development, and access to resources in highland Ecuador', *Economic Geography* 75(4): 395–418.

Bryant, R.L. (1998) 'Power, knowledge and political ecology in the Third World: a review', *Progress in Physical Geography* 22(1): 79–94.

Bumpus, A.G. and Liverman, D.M. (2008) 'Accumulation by decarbonization and the governance of carbon offsets', *Economic Geography* 84(2): 127–155.

Bunce, M. (2008) 'The "leisuring" of rural landscapes in Barbados: new spatialities and the implications for sustainability in small island states', *Geoforum* 39(2): 969–979.

Counsell, D. and Haughton, G. (2006) 'Sustainable development in regional planning: the search for new tools and renewed legitimacy', *Geoforum* 37(6): 921–931.

References

Daily, G.C. and Ehrlich, P.R. (1992) 'Population, sustainability, and Earth's carrying capacity', *BioScience* 42: 761–771.

Daly, H.E. (1977) *Steady-State Economics: The Economics of Biophysical Equilibrium and Moral Growth*. San Francisco, CA: W.H. Freeman.

Daly, H. and Farley, J. (2003) *Ecological Economics: Principles and Applications*. Washington, DC: Island Press.

Dresner, S. (2008) *The Principles of Sustainability*, 2nd edn. London: Earthscan.

Frosch, R.A. (1995) 'Industrial ecology: adapting technology for a sustainable world', *Environment* 37(10): 16–28.

Gibbs, D. (2003) 'Trust and networking in inter-firm relations: the case of eco-industrial development', *Local Economy* 18(3): 222–236.

Goldman, M. (2004) 'Eco-governmentality and other transnational practices of a "green" World Bank', in J.R. Peet and M. Watts (eds) *Liberation Ecologies*, 2nd edn. London: Routledge. pp. 166–192.

Grossman, G. and Krueger, A.B. (1995) 'Economic growth and the environment', *Quarterly Journal of Economics* 110(2): 353–377.

Hayter, R., Barnes, T.J. and Bradshaw, M.J. (2003) 'Relocating resource peripheries to the core of economic geography's theorizing: rationale and agenda', *Area* 35(1): 15–23.

Huber, J. (2000) 'Towards industrial ecology: sustainable development as a concept of ecological modernization', *Journal of Environmental Policy and Planning* 2: 269–285.

Krueger, R. and Gibbs, D. (eds) (2007) *The Sustainable Development Paradox: Urban Political Economy in the United States and Europe*. New York: Guilford Press.

Leichenko, R. and O'Brien, K. (2008) *Environmental Change and Globalization: Double Exposure*. Oxford: Oxford University Press.

Leichenko, R.M. and Solecki, W.D. (2005) 'Exporting the American dream: the globalization of suburban consumption landscapes', *Regional Studies* 39(2): 241–253.

Maxey, L. (2006) 'Can we sustain sustainable agriculture? Learning from small-scale producer-suppliers in Canada and the UK', *Geographical Journal* 172(3): 230–244.

McCarthy, L. (2002) 'The brownfield dual land-use policy challenge: reducing barriers to private redevelopment while connecting reuse to broader community goals', *Land Use Policy* 19(4): 287–296.

McGregor, A. (2004) 'Sustainable development and "warm fuzzy feelings": discourse and nature within Australian environmental imaginaries', *Geoforum* 35(5): 593–606.

McManus, P. and Gibbs, D. (2008) 'Industrial ecosystems? The use of tropes in the literature of industrial ecology and eco-industrial parks', *Progress in Human Geography* 32(4): 525–540.

O'Brien, K. and Leichenko, R. (2006) 'Climate change, equity and human security', *Erde* 137(3): 165–179.

Pacione, M. (2007) 'Sustainable urban development in the UK: rhetoric or reality?', *Geography* 92(3): 248–265.

Pellow, D.N. (2007) *Resisting Global Toxics: Transnational Movements for Environmental Justice*. Boston: MIT Press.

Ponte, S. (2008) 'Greener than thou: the political economy of fish ecolabeling and Its local manifestations in South Africa', *World Development* 36(1): 159–175.

References

Redclift, M. (1993) 'Sustainable development: needs, values, rights', *Environmental Values* 2: 3–20.

Redclift, M. (2005) 'Sustainable development (1987–2005): an oxymoron comes of age', *Sustainable Development* 13: 212–217.

Robbins, P. (2004) *Political Ecology: A Critical Introduction*. Malden, MA: Wiley-Blackwell.

Rock, M.T. and Angel, D.P. (2006) *Industrial Transformation in the Developing World*. Oxford: Oxford University Press.

Sneddon, C., Howarth, R.B. and Norgaard, R.B. (2006) 'Sustainable development in a post-Brundtland world', *Ecological Economics* 57(2): 253–268.

Soyez, D. and Schulz, C. (2008) 'Facets of an emerging Environmental Economic Geography (EEG)', *Geoforum* 39(1): 17–19.

Tietenberg, T. (2006) *Environmental and Natural Resource Economics*, 5th edn. Reading, MA: Addison-Wesley.

Wackernagel, M. and Rees, W. (1996) *Our Ecological Footprint: Reducing Human Impact on the Earth*. Gabriola Island, BC: New Society Publishers.

World Bank (2003) *World Development Report 2003: Sustainable Development in a Dynamic World*. Washington, DC: World Bank.

World Commission on Environment and Development (WCED) (1987) *Our Common Future*. Oxford: Oxford University Press.

Yohe, G. and Schlesinger, M. (2002) 'The economic geography of the impacts of climate change', *Journal of Economic Geography* 2(3): 311–341.

INDEX

Index

278